Grade 4 • Unit 4

Inspire Science

Information Processing and Living Things

FRONT COVER: (t)agustavop/iStock/Getty Images, (c)Ken Cavanagh/McGraw-Hill Education, (b inset)Alexander Raths/Shutterstock; **BACK COVER:** agustavop/iStock/Getty Images

Mheducation.com/prek-12

STEM McGraw-Hill is committed to providing instructional materials in Science, Technology, Engineering, and Mathematics (STEM) that give all students a solid foundation, one that prepares them for college and careers in the 21st century.

Send all inquiries to:
McGraw-Hill Education
8787 Orion Place
Columbus, OH 43240

ISBN: 978-0-07-699637-7
MHID: 0-07-699637-9

Printed in the United States of America.

7 8 9 10 11 LWI 26 25 24 23 22 21

Table of Contents
Unit 4: Information Processing and Living Things

Structures and Functions of Living Things

Information Processing and Transfer

Structures and Functions of Living Things

ENCOUNTER THE PHENOMENON

How do structures help living things survive?

GO ONLINE
Check out *Butterfly* to see the phenomenon in action.

Talk About It

Look at the photo and watch the video of monarch butterflies. What structures do you observe? What are you curious about? Talk about your thoughts with a partner.

Did You Know?

Monarch butterflies have organs located on their feet and heads that help them identify different plants.

STEM Module Project Launch
Science Challenge

National Park Presentation

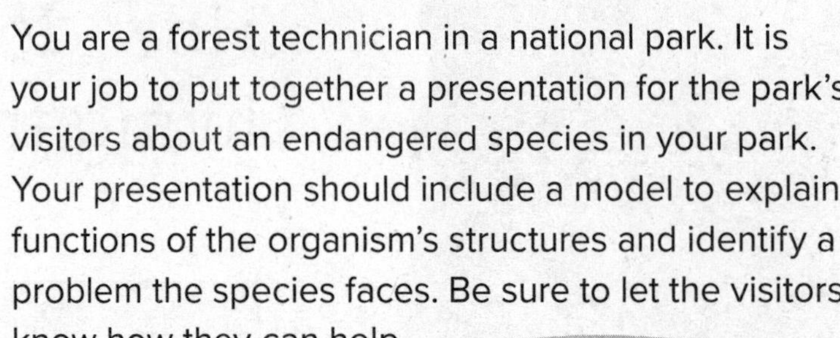

Lesson 1
Structures and Functions of Plants

You are a forest technician in a national park. It is your job to put together a presentation for the park's visitors about an endangered species in your park. Your presentation should include a model to explain functions of the organism's structures and identify a problem the species faces. Be sure to let the visitors know how they can help.

Congratulations! As a forest technician, you will teach visitors how an organism's structures help it survive.

Lesson 2
Structures and Functions of Animals

Do you enjoy talking to people and being outside? Forest technicians work outside, protecting the woodlands. They also help educate visitors about a park's wildlife and natural history.

POPPY
Park Ranger

STEM Module Project

Plan and Complete the Science Challenge
You will use what you learn to teach others about how an organism's structures help it survive.

LESSON 1 LAUNCH

Plant Parts

Plants are made up of different parts that help a plant live in its environment. Circle all of the parts that can be found on a plant.

Roots	Leaves	Bark
Flower	Nuts	Seeds
Spines	Stems	Trunk
Branch	Pine needles	Tubes that carry water
Root hairs	Fruit	Waxy coating

Explain your thinking. How did you decide which things were parts of a plant?

You will revisit the Page Keeley Science Probe later in the lesson.

LESSON 1

Structures and Functions of Plants

The coast redwood is a cone-bearing tree found along the coast of the northwest region of the United States.

ENCOUNTER
THE PHENOMENON

Why are these trees so tall?

GO ONLINE

Check out *Forest* to see the phenomenon in action.

Talk About It

Look at the photo and watch the video of the redwood trees. Circle items in the photo that you can compare to the trees to judge their height. What do you observe? Talk about your observations with a partner. Record or illustrate your thoughts below.

Did You Know?

Redwood trees can grow to be taller than the Statue of Liberty. They can live to be 2,000 years old.

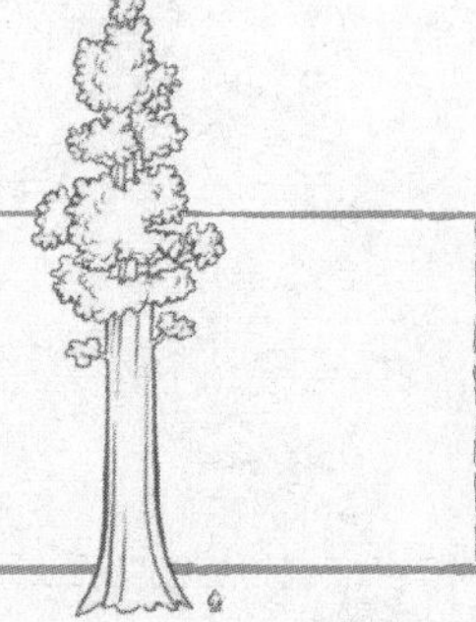

INQUIRY ACTIVITY

Hands On

Plant Parts

Materials

hand lens

Think of all the plants that you observed on your way to school. Earth has a great variety of plants. All plants are made of structures, or parts, each having a specific purpose. In this investigation, you will be comparing two different plants.

Make a Prediction How do the same plant parts compare between two different plants?

Carry Out an Investigation

BE CAREFUL Observe the plants without touching them.

1. Take a walk around your schoolyard with your class.
2. Find two different plants to observe and sketch. Draw the plants below. Use the hand lens to get a better look at the plants' parts.
3. Identify as many of the plant parts as you can. Label them in your drawings.

Plant 1	Plant 2

Communicate Information

4. How were the shapes of the leaves on the two different plants similar and different?

5. How were the shapes of the stems on the two different plants similar and different?

6. Think about the plants that you observed and other plants that you have seen. How are plants different from one another?

Talk About It

Did your observations support your prediction? Discuss with a partner.

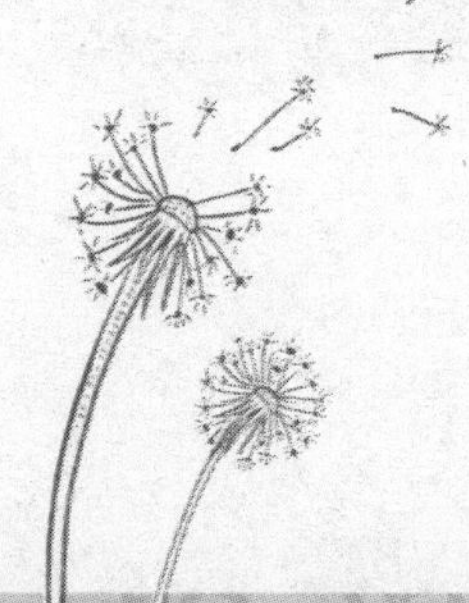

VOCABULARY

Look for these words as you read:

adaptation

response

stimulus

transpiration

tropism

Plant Needs

The redwood trees that you saw in the lesson phenomenon and the two schoolyard plants you observed in the Inquiry Activity, *Plant Parts,* probably look very different. But they aren't as different as you might think. All plants have the same basic needs and a set of typical structures.

The basic needs of plants are air, water, sunlight, nutrients, and space. Plants must live in an environment where their needs are met.

The air around Earth is a mixture of gases. Plants need one of these gases, carbon dioxide, to make food. They need another gas, oxygen, to break down the food. Plants have pores, or stomata, in their leaves that allow gases to move in and out of the plant.

Plants use sunlight to make food. They use the energy from sunlight to make sugar. The sugar provides the energy plants need to survive. Some plants need more sunlight than others. Plants use their leaves to gather sunlight.

Palm trees require a lot of sunlight. Mosses and ferns can grow in shady areas.

All living things, including plants, need water. Water is another material that plants use to make food. Water is also used to move nutrients through plants. In plants, water also provides support. Plants that do not get enough water will start to droop, or wilt. Most plants take in water through their roots. The water then moves into a system of tubes that distribute the water throughout the plant.

Substances that a living thing needs to stay healthy are called nutrients. Plants need nutrients found in their environment. Most plants take in nutrients, which are dissolved in water, through their roots.

Plants need enough space to get the air, water, sunlight, and nutrients they need to survive. Plants that are crowded close together have a harder time getting the things they need.

Some plants need more water than others. Cacti can survive in deserts with little rain, while the plants in a rainforest prefer a very wet area.

What needs might not be met if you plant the plants in your garden too close together?

Plant Parts

Most plants have roots, stems, and leaves. These parts, or structures, help the plant meet its needs and carry out life functions.

Roots

Plant roots take in water and dissolved nutrients from the soil. Roots also hold the plant in place. Some roots store food the plant has made.

Stem

The stem supports the plant. It is also part of a plant's transport system. There are two types of stems: soft stems and woody stems. Soft stems are green and are flexible. Woody stems are hard and are often covered in bark. Tree trunks are examples of woody stems.

Stems also allow materials to move inside the plant through a system of tubes. The tubes in the stems carry water and dissolved nutrients.

Label a Diagram: Roots, Stems, and Leaves

Label the different parts of the plant. Then describe the functions of each of the plant parts below.

Roots:

Stems:

Leaves:

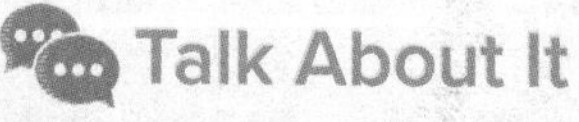

Talk About It

Use evidence to construct an argument that plant structures function to support survival.

Leaves

Leaves vary in shape and size. Most leaves have a line of symmetry, which means their shape can be divided into two identical parts. Leaves are the site of gas exchange and food production. Most leaves are broad and flat, which allows them to collect sunlight efficiently. Plants break down the food they make in the leaves and use it for growth and repair. **Transpiration** is the release of water vapor, mainly through the small openings in the underside of leaves. This process drives the movement of material throughout a plant.

GO ONLINE Watch the video *Plant Structure and Function* to see more plant structures and functions.

Talk About It

What parts of a plant have a line of symmetry? Discuss with a partner.

FOLDABLES®

Cut out the Notebook Foldables tabs given to you by your teacher. Glue the anchor tabs. Use evidence to explain how the structures present in the photo help plants survive and grow.

Glue anchor tab here.

Plant Reproduction

Plants have many structures that are used for reproduction. Some of the structures are flowers, cones, seeds, and fruits.

Flowers

Most flowers contain male and female parts. The stamen is the male part. It contains the anther, where pollen is produced. The pistil is the female part. It contains the ovary, where egg cells are produced. Insects, birds, and wind help move pollen. Fertilization occurs once the content inside the pollen joins the egg cells inside the ovary. Seeds develop after fertilization.

Although flowers come in different colors and shapes, they all contain the same structures used in reproduction.

Cones

Some seed plants reproduce with cones. These plants usually produce both male and female cones. The male cones produce pollen that is released into the wind. The female cones produce a sticky liquid that captures the pollen. Fertilization occurs in the female cone.

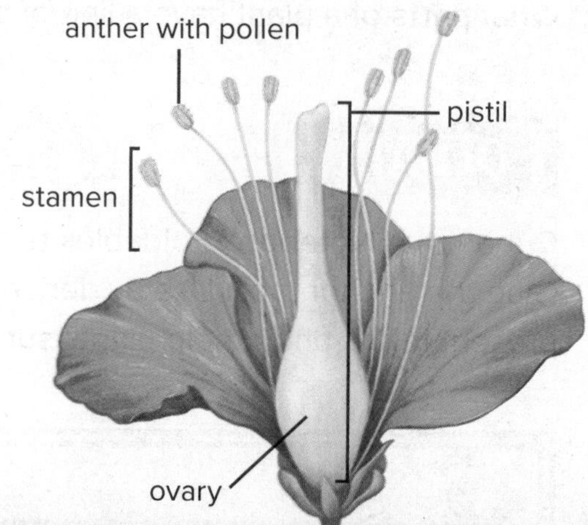

Talk About It

Explain to a classmate the parts of a flower that help it reproduce.

Pollen is produced in the smaller male cones. Seeds develop in the larger female cones.

Seeds

A seed contains an embryo surrounded by a food supply, or cotyledon, and an outer seed coat. An embryo is the beginning of a new organism. It will live off of the food supply until it is big enough to make its own food.

Fruits

As a seed develops, the ovary enlarges into a fruit, which protects the seed. Some fruits appeal to animals, which eat the fruits and spread the plant's seed in their droppings.

Plants rely on animals to scatter their seeds.

Seed coat
Cotyledon
Embryo

Use evidence to construct an argument that a plant's system has parts that work together to support reproduction.

REVISIT Revisit the Page Keeley Science Probe on page 5.

Plant Survival and Behavior

Environments can present challenges to the organisms that live there. An **adaptation** is a physical trait or behavior that helps an organism survive in its environment.

GO ONLINE Use the simulation *Plant Structures* to learn how the structures in plants function.

Many plants have parts that are physical adaptations. For example, desert plants have adaptations for living in a hot, dry environment. Cacti have thick, waxy stems that store water. They have dense, shallow roots to soak up rain quickly. Rainforest plants, such as orchids, have adaptations that help them survive in hot, wet conditions. An orchid's aerial roots absorb nutrients and anchor the plant high in a tree. Orchids also have leaves that are shaped to drain excess water to prevent rotting.

Many plants have adaptations to defend themselves from animals that would eat them. Some plants have thorns. Others produce chemicals that are poisonous or taste bad.

Construct an argument from evidence to explain how the **parts** of each plant's **system** help it survive.

Cacti store water in their stems.

Orchids have aerial roots and drip tip leaves that help them survive in warm, moist conditions.

Plant Behavior

Plants have internal structures that enable them to react to changes in their environments. A change in an environment that causes an organism to respond is called a **stimulus**. The reaction or change in behavior of an organism is called a **response**. Plants respond to stimuli such as sunlight, water, and gravity.

A plant responds to a stimulus by changing its pattern of growth. A plant's response to water, gravity, light, and touch is called **tropism**.

Plants respond to light by growing toward the light source. This response is known as phototropism. Most plant roots grow downward, the same direction as the pull of gravity, while most stems grow upward. This is called gravitropism. Roots sense water in the soil and grow toward or away from it. This response is known as hydrotropism. Some plants respond to touch, or contact with an object, by curling around that object or clinging to it. This is known as thigmotropism.

1. List the types of tropisms on the lines below. Identify and label the types of tropism shown in the photos on this page.

2. Use evidence to construct an argument that plants have structures that support behavior.

STEM Connection

How Could You Become a Horticultural Scientist?

Do you like growing plants and helping people? **Horticultural scientists** help farmers improve crop production. Plants are an essential part of our lives. All of the food that we eat comes directly or indirectly from plants. Horticultural scientists play an important role in making sure we have enough food to eat. Some horticultural scientists conduct research. Others work with farmers to help them grow more plants. They have knowledge of plant biology, soils, and pests that might affect crops. Some study genetics and diseases.

To become a horticultural scientist, you will take courses in biology, botany, soil science, and entomology, the study of insects.

It's Your Turn

Think like a horticultural scientist as you study plants in different habitats in the next activity.

Talk About It

How would knowing about different habitats help a horticultural scientist?

INQUIRY ACTIVITY

Research

Survival in Different Habitats

The plants that you observed in the Inquiry Activity, *Plant Parts*, all live in the same habitat, yet their structures are different. There are many different habitats found throughout the United States. How will the structures of plants that live in different habitats differ?

Make a Prediction Choose two habitats from the table below. Make a prediction about how plants from the two habitats will differ.

Carry Out an Investigation

1. Use the table below to choose two plants from different habitats.

Type of Habitat	Common Plants
Tundra	arctic moss, reindeer lichen
Northwestern Forested Mountains	thimbleberry
Eastern Temperate Forests	spruce, magnolia
Great Plains	big bluestem grass, buffalo grass
North American Desert	saguaro cactus, desert holly
Northwestern Coastal Forests	coast redwood

INQUIRY ACTIVITY

2. Research the two habitats from which you chose your plants. Describe the habitats below.

3. Research each plant you chose. Identify and describe its structures, including any adaptations. Draw the plants and label their structures below.

4. What adaptations are found in each of the plants that you chose?

5. **ENVIRONMENTAL Connection** Do more research to find out how humans have affected each of the habitats that you chose. Take notes below.

Communicate Information

6. How did the plants that you chose for your investigation differ?

Engage in an argument from evidence to explain how the plants from your investigation have **structures that function** to support survival in different environments.

Talk About It

How is an organism's structure related to its function? Discuss your ideas with a partner.

EXPLAIN THE PHENOMENON

Why are these trees so tall?

Summarize It

Construct an argument about how structure in plants support growth, survival, and reproduction.

REVISIT

Revisit the Page Keeley Science Probe on page 5. Has your thinking changed? If so, explain how it has changed.

Three-Dimensional Thinking

1. What do plants need to survive?

A. Nutrients

B. Sunlight

C. Gases

D. All of the above

2. A(n) ____________________ is a physical trait or behavior that helps an organism survive.

3. How is structure related to function?

Extend It

How does human activity affect plants? Think back to the Inquiry Activity, *Survival in Different Habitats*. Identify a negative effect that human activity has on plants, and propose a solution to the problem.

OPEN INQUIRY

What questions do you still have about plants' structures and their functions?

Plan and carry out an investigation or research to find the answer to your question.

KEEP PLANNING

STEM Module Project

Science Challenge

Now that you have learned about the structures and functions of plants, go to your Module Project to explain how this information will affect your plan for your national park presentation.

LESSON 2 LAUNCH

Animal Parts

Animals are made up of different parts that help an animal live in its environment. Circle all of the parts that could be found on an animal.

Ear	**Shell**	**Claw**
Heart	**Leaf**	**Feather**
Tentacle	**Tail**	**Fur**
Roots	**Lungs**	**Skin**
Teeth	**Antennae**	**Wings**
Fin	**Beak**	**Seeds**

Explain your thinking. How did you decide which things are parts of an animal?

____

____

____

____

You will revisit the Page Keeley Science Probe later in the lesson.

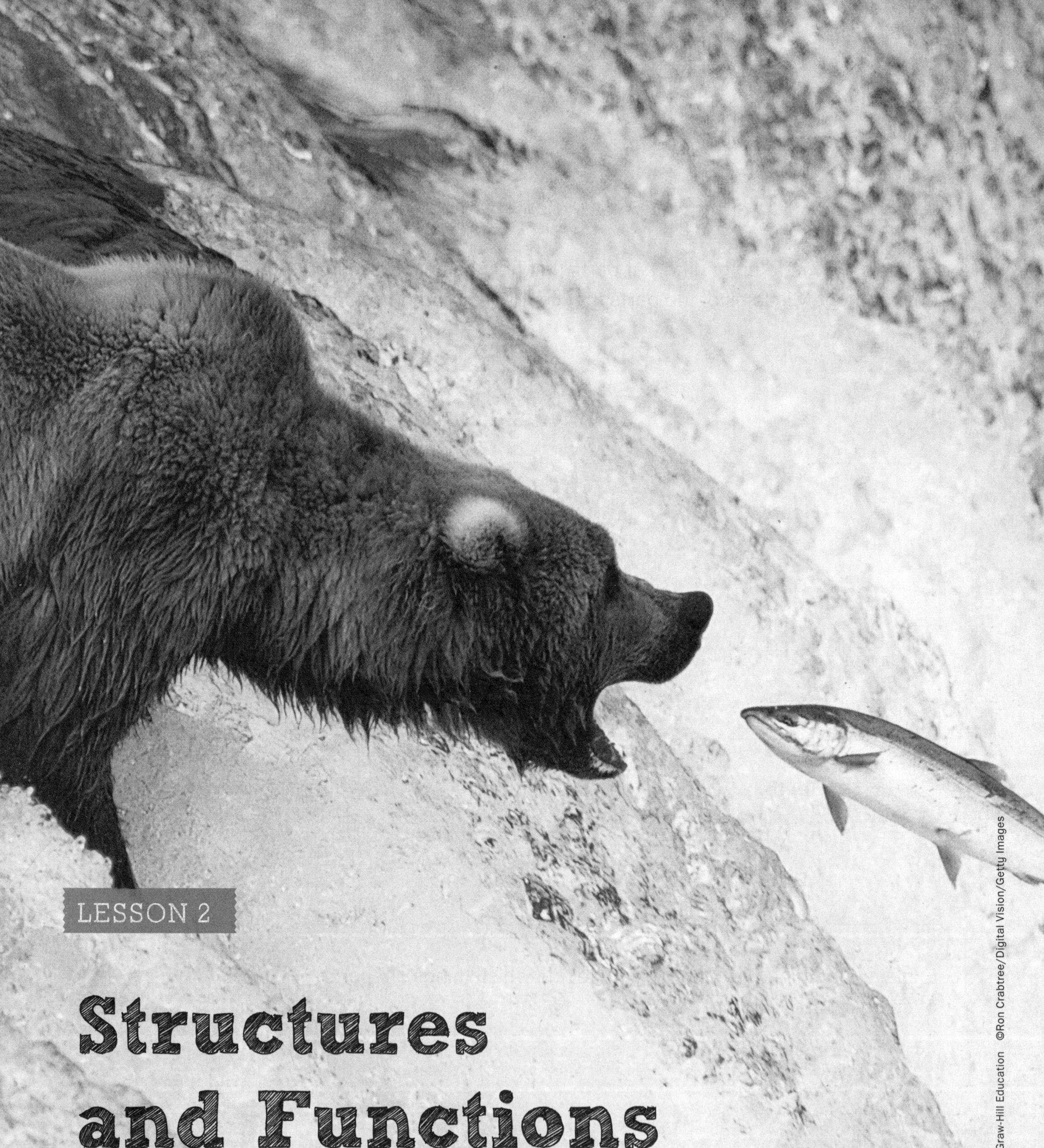

LESSON 2

Structures and Functions of Animals

ENCOUNTER
THE PHENOMENON

How do grizzly bears use their structures?

GO ONLINE

Check out *Grizzly Bear* to see this phenomenon in action.

Look at the photo and watch the video of grizzly bears. What do you observe? Discuss your thoughts with a partner. Record your thoughts below.

Did You Know?

Grizzly bear claws are 5 to 10 centimeters (2–4 inches) long.

INQUIRY ACTIVITY

Hands On

Animal Parts

Materials

colored pencils

There is a great variety of life on Earth. Apart from plants, this planet is crawling with life, from the tiny earthworm to the elusive mountain lion that roams the mountains. Think of all the animals that you hear or see when you take a walk outside. In this investigation, you will compare the parts, or structures, of two different animals.

Make a Prediction How do the same structures on two different animals compare?

Carry Out an Investigation

1. Think of two different animals you saw.
2. Draw the first animal you chose, and label all the parts of the animal that you know.

3. Draw the second animal you chose, and label its parts.

4. In the table, list each animal structure that you labeled. Think about how the animals use each structure, and add the possible function to the table. Use a separate piece of paper if needed.

Structure	Possible Function

INQUIRY ACTIVITY

Communicate Information

5. Think back to the module phenomenon of a bear. How is one of the animals you chose similar to a bear? How are they different?

6. How are the structures of the two animals you chose alike?

7. How are the structures of the two animals you chose different?

Talk About It

Compare one of your animals to one of a classmate's animals. How are their structures alike? How are they different? Talk to your partner.

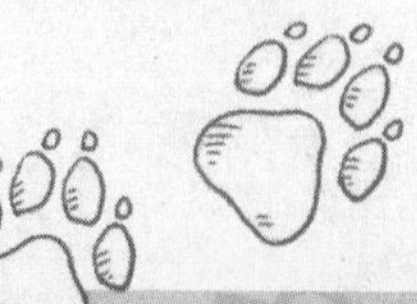

MAKE YOUR CLAIM

Think back to the different animals in the investigation. How do animals use their structures to support survival, growth, behavior, and reproduction?

Make your claim. Use your investigation.

CLAIM

Animals have external and internal structures that ______________________ to support survival, growth, behavior, and reproduction.

Cite evidence from the activity.

EVIDENCE

The investigation showed that ______________________.

Discuss your reasoning as a class. Tell about your discussion.

REASONING

The evidence supports the claim because ______________________.

You will revisit your claim to add more evidence later in this lesson.

Animal Needs and Structures

VOCABULARY

Look for these words as you read:

external structure

internal structure

structural adaptation

Animals need food, water, space, and shelter to survive. While all animals have the same needs, they meet their needs in a variety of ways and places.

Food, Oxygen, and Water

Unlike plants, which produce their own food, animals eat other organisms to get energy. Animals use the energy from food to grow, survive, and reproduce. Oxygen, a gas found in air and water, helps animals get energy from food. Water keeps the parts inside an animal's body working properly.

Some animals that live in water need to come to the surface to breathe air. However, most animals that live in water take in oxygen from the water. For example, fish are adapted to exchange gases with the water that surrounds them. They have structures called gills for this purpose.

Butterflies have a body part called a proboscis which helps them get food from flowers and fruits.

Space and Shelter

Animals need space to hunt for food, escape from predators, and build homes. A shelter is where an animal makes its home. Shelters provide protection for the animals that live in them. They also provide a place for animals to give birth and raise their young. Birds build nests in trees and on cliffs. Woodchucks and squirrels dig burrows.

1. Explain why animals need food and space.

An abundant water supply is important for all living things.

Structures

Structures inside and outside animals' bodies work together to obtain nutrients, digest food, eliminate waste, and reproduce. These parts keep an animal alive and help it reproduce.

GO ONLINE Watch the video *Animal Structures* to learn more about animal structures.

Internal structures are structures found inside an organism's body. These structures, like the major organs, have specific functions. For example, the brain's main function is to process information. The stomach helps digest food. The intestines absorb nutrients. Kidneys help eliminate waste. Lungs help with the exchange of gases. Animals can reproduce once their reproductive organs reach maturity.

Orangutans move by swinging from tree to tree. They need plenty of space to find shelter and food.

External structures are found outside of an organism's body. A shark's teeth and a bird's beak are examples of external structures that help these animals get food.

2. **WRITING Connection** Revisit the Explore activity. Research the two different animals that you compared, focusing on their internal structures this time. Write a short summary explaining how these structures help each animal survive its environment. Use a separate piece of paper if needed.

Robins build nests that are high off the ground so that they can safely lay eggs and raise their young.

Fish take in oxygen from the water through their gills.

Structural Adaptations

An organism's **structural adaptations** are inherited changes to physical features that help an organism survive and reproduce. Fur color, long limbs, strong jaws, and the ability to run fast are structural adaptations. Some structural adaptations help organisms survive in certain environments. Other structural adaptations protect prey from predators or enable predators to hunt more successfully.

Camouflage

Camouflage is any coloring, shape, or pattern that allows an organism to blend in with its environment. Predators with camouflage sneak up on prey. Camouflage also helps prey animals hide from predators.

Mimicry

Mimicry is an adaptation in which an animal is protected against its predators by its resemblance to a different animal or object. For example, the spicebush swallowtail caterpillar's head has spots that look like a snake's head. This shape frightens away most predators.

This stick bug avoids predators by looking just like bark, leaves, or twigs.

Spicebush swallowtail caterpillar

1. Explain how structural adaptations help animals survive.

____

____

____

____

____

2. Circle the animal that shows mimicry. Put a square around the animal that is using camouflage.

REVISIT Revisit the Page Keeley Science Probe on page 25.

PAGE KEELEY SCIENCE PROBES

Behavioral Adaptations

An adjustment in an organism's behavior is a behavioral adaptation. For example, many animals travel in herds for protection from predators. Others, such as wolves, hunt in packs to capture larger prey.

Many animals—such as birds, butterflies, and fish—migrate. Migration is the movement of animals from one place to another. Animals migrate to find food, reproduce in better conditions, or find a less severe climate.

Some animals endure cold winters by hibernating. Hibernation is a period of inactivity during cold weather. During this time, animals remain inactive until warmer temperatures return in spring. Grizzly bears prepare for hibernation around November. They will not eat, drink, or eliminate bodily wastes during hibernation, which lasts approximately five months.

GO ONLINE Explore *How Animals Survive* to learn more about what helps animals stay alive.

The desert kangaroo rat is adapted to get all of the water it needs from food. It never needs to drink.

1. Why do animals travel in herds?

2. What is the advantage of hunting in a group?

Talk About It

Why do animals hibernate during winter and not during summer? Discuss with a partner.

Inspect

Read the passage *Winged Mysteries*. Underline in the passage the evidence that shows how butterflies survive the cold weather.

Find Evidence

Reread the passage. Discuss with a partner. Why are monarch butterflies "Winged Mysteries?" Highlight text that helps you make an inference.

Notes

Winged Mysteries

Sometime this year, mysterious, six-legged creatures may land in your back yard. They're seeking a place to rest from their long journey—or searching for food. Maybe you've never thought of monarch butterflies as mysterious creatures. But their migration is so mysterious that entomologist Lincoln Brower has spent over 40 years trying to unlock their secrets.

Monarchs are tropical butterflies that must escape a freezing winter to survive. For those that live in southern Canada, it is a long trip to warmer weather. In late August, hundreds of millions of monarchs head south. How can such seemingly fragile insects fly so far? "Monarchs are really good glider pilots," explains Brower. "One of the reasons they get to Mexico without getting their wings damaged is that they're not power flying—they're taking advantage of cold fronts." The monarchs will catch a breeze of cold air and glide during the day, and rest at dusk.

Monarchs are huddled together in their winter home.

After the morning sun warms the butterflies, they take off one by one to catch the rising warm air, and then continue gliding south.

By November, the monarchs have made it to their winter home. They have never been here before but many generations before them have been. Fir forests in Central Mexico and groves of trees in California are blanketed in monarchs. The cool air keeps the butterflies inactive. In the spring, they mature and mate. After the females lay eggs the parent butterflies die. It's up to the next generation to continue the journey back north.

Make Connections

Talk About It

Discuss with a partner why a monarch butterfly migrates instead of hibernates. Look back at pages earlier in the lesson to be reminded of the similarities and differences for both.

Notes

INQUIRY ACTIVITY

Hands On

Put Your Best Foot Forward

You have read that animals have different structures that help them to be better suited to their environment. Have you ever wondered what type of feet would be best for swimming? What type of feet would be best for picking things up? Look at the tongue depressor, fork, and tweezers. You will use these tools to simulate how birds use their feet to survive in nature.

Make a Prediction Take a look at the activites listed in the data table. Which tool will be best for each activity?

____

____

____

____

____

____

Carry Out an Investigation

BE CAREFUL Wear safety goggles during this activity.

1. Add water to one container until it is half full. Add about 2 inches of pea gravel to the second container.
2. Place a handful of the colored gravel into each container.
3. Observe the different tools. They represent different types of feet you might find on birds.
4. Experiment with the tools to determine which would best help a bird to grasp and pick up food (colored gravel), swim in water, and dig through debris (pea gravel).
5. **Record Data** In the table, rate each tool from one to three based on how well it performed in each test. One is good, two is ok, and three is poor.

Materials

safety goggles

2 plastic containers

water

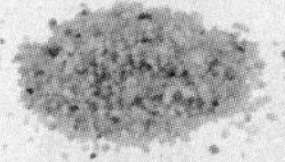
pea gravel

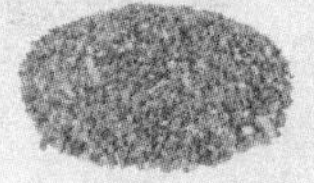
colored gravel

tongue depressor

fork

tweezers

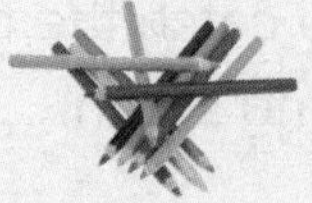
colored pencils

Types of Activity			
	Digging for food in debris	Paddling through water	Picking up and grasping food
Tongue depressor			
Fork			
Tweezers			

6. **Analyze Data** In the data table, use orange to shade the tool that was best for grasping and picking up. Use blue to shade the tool that was best for paddling in water. Use green to shade the tool that was best for digging.

Communicate Information

7. Why do you think it is helpful for birds to have different types of body structures?

Use evidence **to engage in an argument** to support a claim about the **function of structures** in different birds.

Talk About It

Did the result support your prediction? Discuss your results with a partner.

COLLECT EVIDENCE

Add evidence to your claim on page 31 about how structures work together to help animals survive.

STEM Connection

What Does a Zoologist Do?

A zoologist is a scientist who studies animals. Zoologists also study animals that are extinct. Zoologists observe animals in their natural habitat and in a laboratory and collect data to learn more about the animals they are studying. They study an animal's physical characteristics, diet, and behaviors. They often work for universities, museums, or zoos. They may work outside in the field, in a laboratory, or in an office.

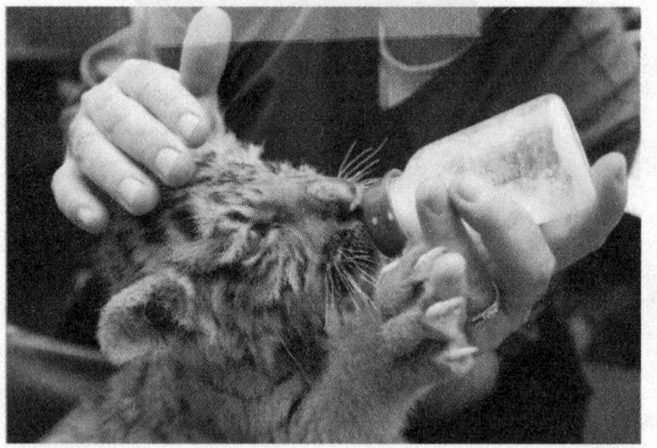

It's Your Turn

Like a zoologist, you will now analyze data to learn more about animals. Complete the activity on the next page to find out more about how birds' structures help them survive in their environment.

Talk About It

How might a zoologist help an animal survive? Discuss with a partner.

Darwin's Finches

Charles Darwin was a British naturalist, geologist, and biologist who studied the variation in structures of living things. In 1831, he traveled around the world for five years. During his voyage, he arrived at the Galápagos Islands and started to explore. He noticed that each island had different types of food available to finches. The food supply included thick-shelled seeds, insects, fruits, and other resources.

PRIMARY SOURCE

Analyze the sketches of the finches' beaks. Form an argument, supported by evidence, about how the finches' beaks allow them to obtain food from their environment.

EXPLAIN THE PHENOMENON

How do grizzly bears use their structures?

Summarize It

Explain the function of structures in animals.

REVISIT Revisit Page Keeley Science Probe on page 25. Has your thinking changed? If so, explain how it has changed.

Three-Dimensional Thinking

1. Animals need certain things to survive. Circle all of the things animals need to survive.

   A. Animals need food for energy.

   B. Animals need water to keep their organs working properly.

   C. Animals need sunlight to make their own food.

   D. Animals need oxygen to raise their young.

2. Use an analogy to help explain how internal and external body structures work together to help an animal stay alive.

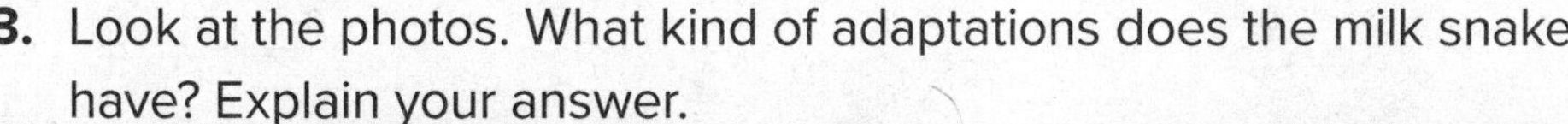

3. Look at the photos. What kind of adaptations does the milk snake have? Explain your answer.

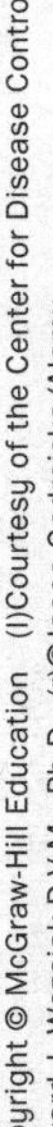

Extend It

You are a zoologist. Study the effects of releasing invasive species into the wild. How might you communicate the effects of invasive species on native species to the people in your community?

Write a speech, draw a poster, create a flyer, or use media.

KEEP PLANNING

STEM Module Project

Science Challenge

Now that you have learned about animals' structures and functions, go to your Module Project to explain how this information will affect your plan for your National Park Presentation.

STEM Module Project
Science Challenge

National Park Presentation

You have been hired as a forest technician in a national park. Using what you have learned throughout this module, prepare a presentation about an endangered species in your park. You will use a model to identify internal and external structures. Include evidence that explains how internal and external structures work together to support the organism's survival, growth, behavior, and reproduction. Be sure to let visitors know about a problem the organism faces and a solution to help save the species.

Planning after Lesson 1

Apply what you have learned about plants and their structures to your project planning.

What factors can decrease the chances of survival in plants?

Record information to help you plan your project after each lesson.

STEM Module Project
Science Challenge

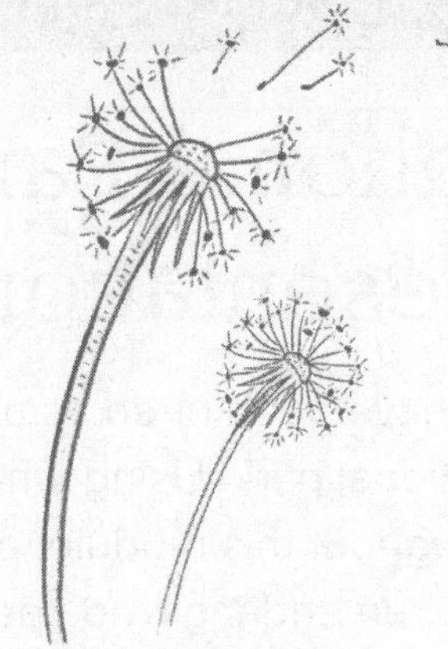

Planning after Lesson 2

Apply what you have learned about the structures found in animals to your project planning.

How can animals survive in a changing environment?

Read the Investigator article, *Artificial Skin*, to find more information about how technology can help people, who have lost a leg or an arm, restore their sense of touch.

How can engineers and doctors help people, who have lost all or part of an arm, feel pressure on their artificial arm?

Research the Problem

Research endangered species by going online to teacher-approved websites, or find books at your local library.

STEM Module Project
Science Challenge

National Park Presentation

Look back at the planning you did after each lesson.
Use that information to complete your final module project.

Define the Problem

Build Your Model

1. Determine the materials you will need to make your presentation and model. List the materials on the lines provided.
2. Use your project planning and your research to design a model of your species' structures.
3. Label the structures on your model.
4. Include relevant and sufficient information about the functions of each structure.
5. Include an explanation of how all the parts work together to support the organism's growth, survival, behavior, and reproduction.
6. Identify a problem the species faces and propose a solution for saving the species.

Materials

Sketch Your Model

Sketch a draft of your model. Remember to label the organism's internal and external structures.

Cite Evidence

Compare your model and research with that of another classmate. Provide feedback about the strengths and weaknesses of their explanation. Determine if they have enough evidence to support their claim of how the structures support survival, growth, behavior, and reproduction. On the lines below, record ideas for improving your own presentation.

Share your plans for your model with another group.

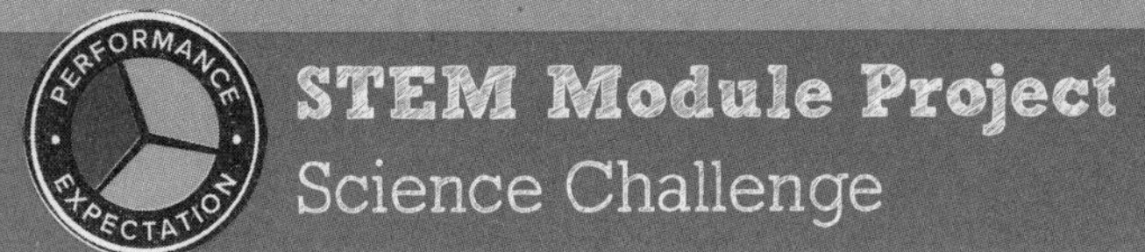

STEM Module Project
Science Challenge

Communicate Your Results

Remember that you will speak as a forest technician when you present your model, along with your research and argument, to the class. Write your research presentation on the lines below.

After listening to all of the presentations, pick one that you like the best. Write an opinion piece on what species is the most important to save. Use evidence to explain your reasoning. Use a separate piece of paper if needed.

MODULE WRAP-UP

REVISIT THE PHENOMENON

Using what you learned in this module, explain how living things use their structures to survive.

Have your ideas changed? Explain your answer.

Information Processing and Transfer

ENCOUNTER
THE PHENOMENON

How does a lighthouse transmit a message across a distance?

GO ONLINE

Check out *Clicks and Echoes* to see the phenomenon in action.

Talk About It

Look at the photo and watch the video. What do you wonder about the phenomenon? Are there any similarities and differences in the way humans and animals transmit and receive information? Talk about what you observed with a partner.

Did You Know?

Lighthouses use a system of lights and lenses to inform boats of dangerous sites along the coastline.

STEM Module Project Launch
Engineering Challenge

Pixel Message

Lesson 1
Information Processing in Animals

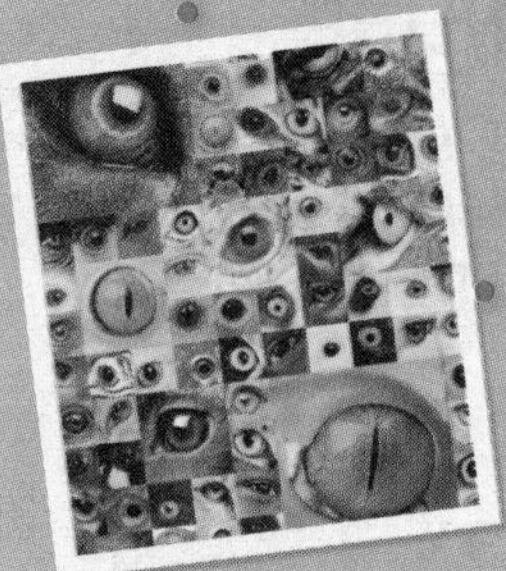

Lesson 2
Role of Animals' Eyes

Lesson 3
Information Transfer

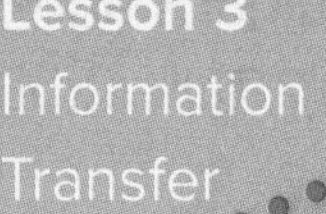

How do we encode messages and send them across the world? Telecommunications engineers design, test, and build technologies that allow information to be transmitted over distances. You are being hired as a telecommunications engineer. It will be your job to design and build a device that uses sound, light, or both, to create two binary codes. Both codes should transmit a pixel message across the classroom to your group. You will compare both binary codes and decide which one is more efficient based on the accuracy and the speed at which the message can be transmitted.

Design a communication device that can transmit a pixel message using two binary codes.

Do you enjoy working with computers? Telecommunications engineers use computers to monitor systems, and they repair any malfunctions.

RUBY
Veterinarian

STEM Module Project

Plan and Complete the Engineering Challenge Use what you learn throughout the module to complete the design of your communication device.

LESSON 1 LAUNCH

Animal Senses

Three friends were talking about how animals use the information they get from their senses. They each had a different idea. This is what they said:

Melinda: *I think each sense organ processes the information from our senses. For example, the nose tells you what the smell is.*

Ralph: *I think the brain processes all the information from our senses. For example, the brain tells you what it is you are seeing.*

Nate: *I think the nerves in our sense organs process the information from our senses. For example, nerves in our fingers tell us what we are touching.*

Whom do you agree with most? ______________________________

Explain why you agree.

____

____

____

____

____

You will revisit the Page Keeley Science Probe later in the lesson.

LESSON 1

Information Processing in Animals

ENCOUNTER
THE PHENOMENON

How do animals learn about their environments?

Check out *Desert Horned Lizard* to see the phenomenon in action.

Talk About It

Look at the photo of the horned lizard. What are you curious about? Share what you observed with a partner. Record your thoughts below.

Did You Know?

Horned lizards are known to puff up their bodies to appear larger when threatened. They also use a variety of gestures to communicate.

INQUIRY ACTIVITY

Hands On

Sense of Touch

Materials

3 sandpaper samples of different grades

material for blindfold

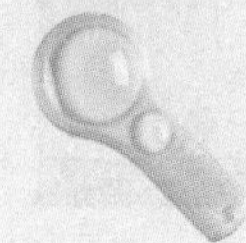

hand lens

How do you react when you touch something hot? Humans, lizards, and other animals use body structures to perceive and respond to their surroundings. In this activity, you will put samples of sandpaper in order by roughness without using your sense of sight. Think about how you use your other senses to obtain information when your eyes are closed.

Make a Prediction How can you use your sense of touch to collect information about your environment?

Carry Out an Investigation

BE CAREFUL Do not move around the classroom with the blindfold on.

1. Have your partner help you put the blindfold on.
2. Your partner will place sandpaper samples textured-side up on a desk in front of you. The samples will be in no particular order.
3. Use your remaining senses to order the samples of sandpaper in a meaningful way.
4. **Record Data** Remove the blindfold. Record the number on the back of each sandpaper sample, and write a description of its texture. Use the data table on the next page.
5. Use the hand lens to observe the size of the grains on each sheet of sandpaper. Describe what you see in the data table.

Sandpaper Order	Description of Texture

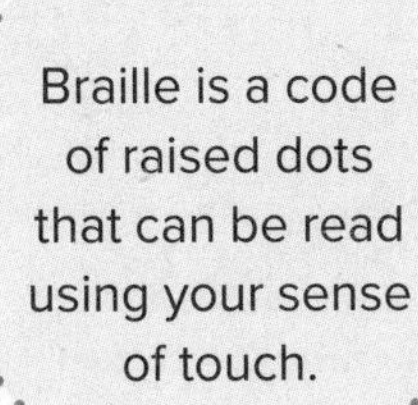

Braille is a code of raised dots that can be read using your sense of touch.

6. Sketch what each sandpaper sample looks like when viewed with the hand lens. Label your sketches.

7. Switch roles with your partner and repeat steps 1–6.

Communicate Information

8. Was your prediction supported by the data you collected? Explain.

9. Compare and contrast the finest and roughest sandpaper.

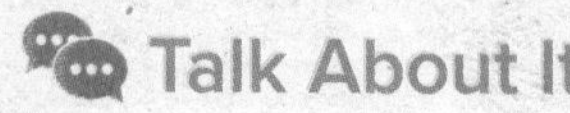

Talk About It

How does the sense of touch help you collect information from your surroundings? In a small group, discuss your ideas.

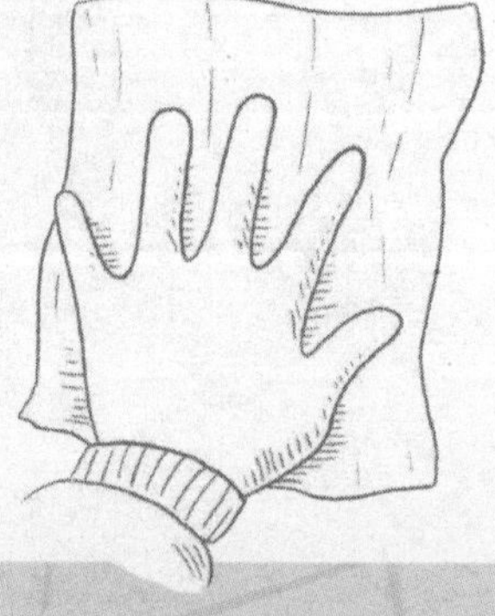

MAKE YOUR CLAIM

How do humans sense and respond to their environments?

Make your claim. Use your investigation.

CLAIM

Humans use their ______________________ to collect information, and respond to their environment.

Cite evidence from the activity.

EVIDENCE

The investigation showed that ________________________.

Discuss your reasoning as a class. Tell about your discussion.

REASONING

The evidence supports the claim because ________________________.

You will revisit your claim to add more evidence later in this lesson.

VOCABULARY

Look for these words as you read:

brain

central nervous system

echolocation

nervous system

peripheral nerve

sensory organ

spinal cord

Animal Senses

Think about how you used your sense of touch in the Inquiry Activity, *Sense of Touch*. Animals also use their senses to learn about their environments. They have **sensory organs,** such as skin, eyes, nose, and tongue, that gather information from outside the body. Like humans, most animals can see, hear, smell, taste, and feel.

Many animals depend on their eyesight to survive. Colossal squids live in ocean waters. They have some of the largest eyes in the animal kingdom. Each eye is about 26 centimeters (10 inches) wide. Large eyes let in a lot of light. This helps the squid see in the dark depths of the ocean.

African elephants have the biggest ears of any animal. Their ears help them hear sounds from very far away. Elephants communicate by making low, rumbling noises that humans cannot hear; however, other elephants can hear them from miles away.

Some animals use different sensory organs than humans. Ants do not have noses. They use their antennae to leave a scent after finding food and to detect smells. Butterflies use their feet to taste leaves. The taste tells them if the plant is a good place to lay their eggs.

1. Underline the sensory organs discussed in the paragraphs above. Circle them in the photos.

2. Draw and label a diagram to show how ants use their sensory organs to learn about their environments.

Jackrabbits have long, pointy ears that give them excellent hearing.

Other Animal Senses

Some animals have senses that humans do not have. Pit vipers and some other snakes have sensory organs that detect infrared light given off by their warm-blooded prey. The light enters a small pit organ, which is located between the snake's eye and nostril. A heat-sensitive part in the organ sends a message to the brain, and the snake strikes.

The duck-billed platypus uses its bill to detect weak electrical fields put out by animals as they move. The bill can also detect movement in the water. The platypus uses this information to quickly locate its prey.

Some bats use echoes to help them navigate and locate prey. **Echolocation** is the process of finding an object by using reflected sound or echoes. These bats send out a high-pitched sound. This sound hits the prey and bounces back to the bat. The bat then interprets this echo to judge the direction and distance of its prey. Some whales and dolphins use echolocation to gather information from their environment.

1. How are the senses of pit vipers different from yours?

2. Use an analogy to explain how echolocation works.

GO ONLINE Watch the video *Animal Senses* to learn about how animals interact with their environment.

The Nervous System

Sensory organs are part of the nervous system. The nervous systems of many animals are similar to the human nervous system. The **nervous system** is the set of organs that uses information from the senses to control all body systems. The **central nervous system** is the part of the nervous system made up of the brain and spinal cord.

The **brain** is an organ that interprets messages received from and sends messages to other body organs. The **spinal cord** is a thick band of nerves inside the spine. It carries messages to and from the brain. Nerves branch off from the spinal cord to all parts of the body. A nerve that is not part of the central nervous system and receives sensory information from other parts of the body is called **peripheral nerve.**

GO ONLINE Explore the *Brain Illumination* simulation to learn more about how your brain interprets stimuli.

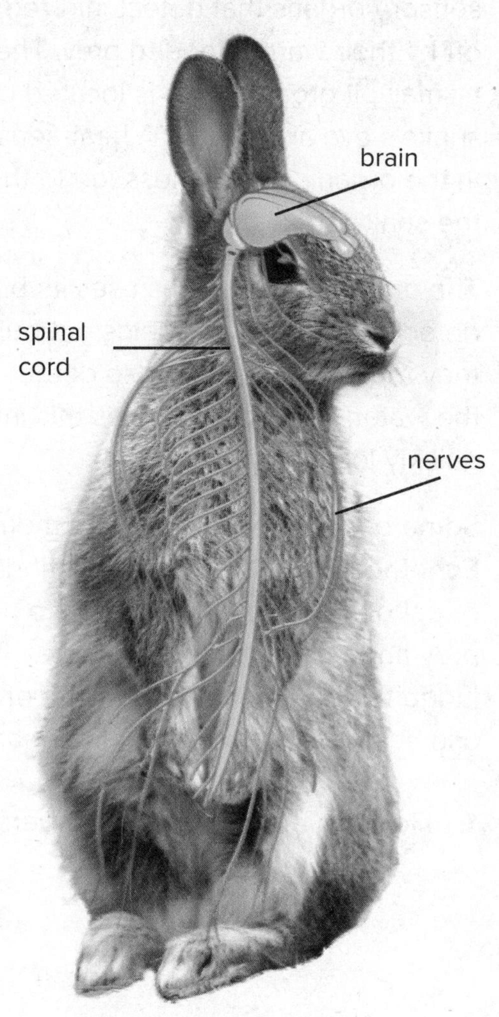

Stimulus and Response

Sensory organs have structures called sensory receptors. Different receptors help detect different types of stimulus from the environment. A stimulus is a thing or event that causes a given response. The process of recognizing and interpreting a stimulus is called perception. The brain then makes sense of the information and tells the body how to respond. Sensory information can be stored as memories that can guide future responses.

1. A rabbit sees a fox. Use the diagram to describe what happens in the rabbit's body that causes the rabbit to run away and hide.

GO ONLINE Explore *The Brain and Nervous System* to learn about systems of information transfer.

Reflexes

The body responds in different ways to stimuli. A reflex is a quick reaction that occurs without waiting for a message to be sent from nerves to the brain. For example, touching something hot causes the hand to quickly pull away. No conscious thought is involved in this response. Instead, this reflex is an action controlled by the spinal cord.

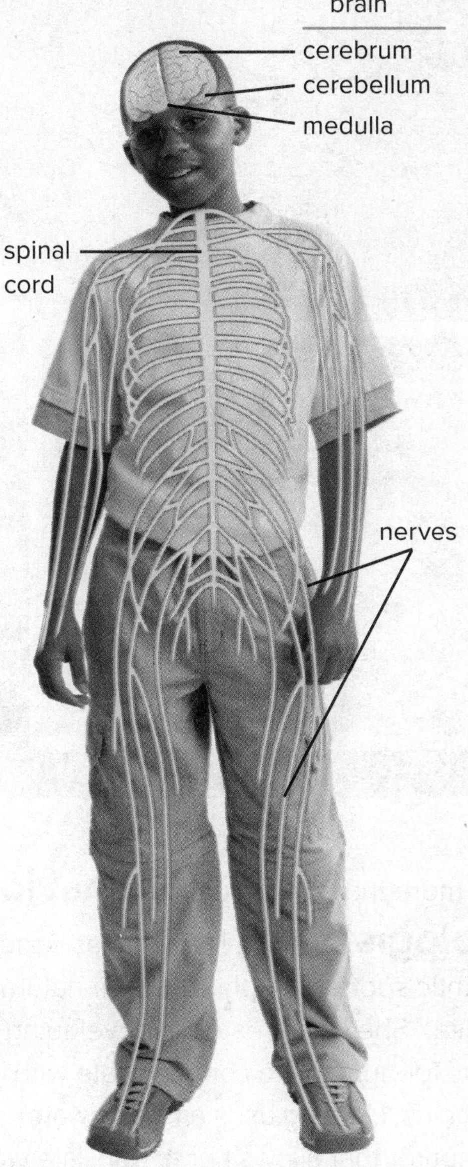

2. **ENGINEERING Connection** Describe a type of technology or tool that has been designed to improve a human sense.

Talk About It

How can memory guide your response to the environment? Discuss with a partner.

People who are visually impaired can process information by using their other senses much more efficiently.

GO ONLINE Use the Personal Tutor *Reflexes* to learn how reflexes work to keep you safe.

REVISIT Revisit the Page Keeley Science Probe on page 55.

STEM Connection

Talk with a Behavioral Biologist

For more than thirty years, **behavioral biologist** Denise Herzing has observed Atlantic spotted dolphins in their natural habitat. She is interested in developing new ways for humans to communicate with dolphins. Herzing uses an underwater computer that allows her to translate words into dolphin whistles and whistles into words.

How do dolphins communicate?

They use a lot of vocalization, but they're very visual also. They have pretty good eyesight, so they use a lot of body postures.

What would you suggest to kids who want to get involved in working with dolphins?

Get as much real-world experience as you can. Volunteer at your local zoo, or in the field with a biologist, counting frogs.

GO ONLINE Explore *Echolocation* to learn how some animals transfer information using sound.

It's Your Turn

Think like a behavioral biologist. Complete the activity on the next page to explore how animals use their senses to react to their environment.

INQUIRY ACTIVITY

Hands On

Pill Bugs

Pill bugs are animals that live under logs, leaves, and rocks. Pill bugs are sometimes called roly-polies, potato bugs, or doodle bugs. They are fascinating creatures. Their bodies are oval and have seven pairs of legs, but they are not insects. They are closely related to shrimp and have gill-like structures to help them breathe. Pill bugs use their senses to detect danger, and they roll into a tight ball when they feel threatened. Now think about what you have learned about senses.

Make a Prediction How will a pill bug react when placed in a dry spot?

____

____

____

____

Materials

4 pill bugs

plastic habitat

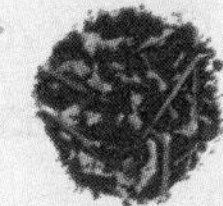
soil with leaves

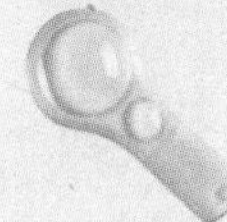
hand lens

paper towels

water

fish food

Carry Out an Investigation

BE CAREFUL Wash your hands before and after handling pill bugs.

1. **Record Data** Place 4 pill bugs in the plastic habitat. Examine the pill bugs with the hand lens. Sketch and label what you see with and without the use of a hand lens. Record observations in the table on the next page.

INQUIRY ACTIVITY

2. Tear two paper towels in half. Make sure the pieces are the same size. Dampen two of the halves.

3. Move the pill bugs to the center of the plastic habitat. Place the moist paper towels in one end of the tray. Place the dry paper towels in the other end.

4. **Record Data** Observe the pill bugs for several minutes. Look for changes in their behavior. Record your observations in the data table.

Time (minutes)	Observations
0	
5	
10	

COLLECT EVIDENCE

Add evidence to your claim on page 61 about how animals process information.

Talk About It

Compare your model to the natural habitat of a pill bug. Discuss with a partner.

Communicate Information

5. Draw a diagram to show how the pill bugs sensed, processed, and responded to the change in their environment.

How could you **use a model** to show how an animal uses its nervous **system** to react to information from the environment?

Talk About It

Did the results support your prediction? Discuss with a partner how the pill bugs perceived their environment when placed in a dry spot.

EXPLAIN THE PHENOMENON | How do animals learn about their environments?

Summarize It

Explain how animals sense, process, and respond to different types of information.

REVISIT

Revisit the Page Keeley Science Probe on page 55. Has your thinking changed? If so, explain how it has changed.

Three-Dimensional Thinking

1. ______________________ is the use of echoes to navigate and to locate prey.

2. The ______________________ form the central nervous system.

   A. peripheral nerves and sensory organs

   B. brain and sensory organs

   C. brain and spinal cord

   D. spinal cord and peripheral nerves

3. Look at the photo below. Explain how information is being processed and transferred.

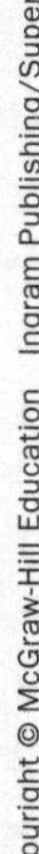

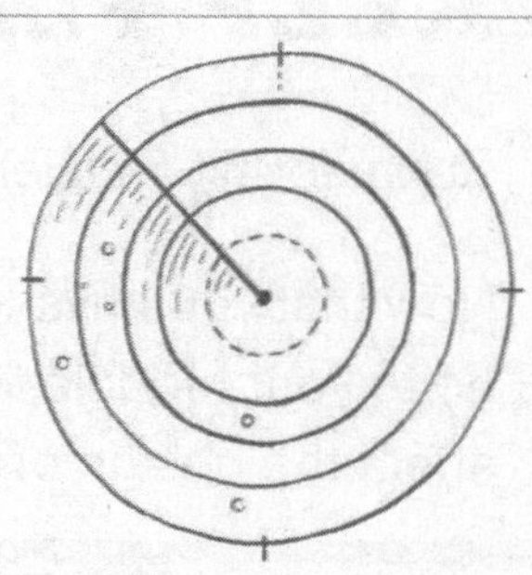

Extend It

You are a behavioral biologist. How might you explain the similar and different ways that humans and animals gather and communicate information to others of the same species?

Write a speech, draw, a poster, create a flyer, or use media.

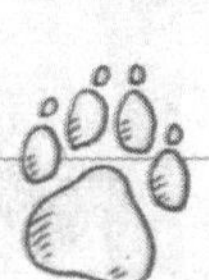

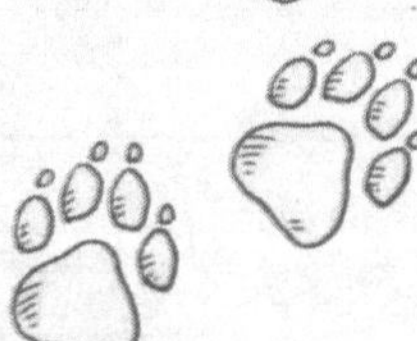

KEEP PLANNING

STEM Module Project

Engineering Challenge

PERFORMANCE EXPECTATION

Now that you have learned about how animals process information, go to your Module Project and explain how this information will affect the design of your communication device.

LESSON 2 LAUNCH

Animal Eyes

Two friends were talking about animals' eyes. They noticed that animals that are active at night have very large eyes. This is what they said:

Laura: *Animals like owls have large eyes so they can see in the dark. They do not need light to see.*

Jayden: *Animals like owls have large eyes to help them see in the dark. They still need some light to see.*

Whom do you agree with more? ______________________________

Explain why you agree.

____

____

____

____

____

____

____

____

You will revisit the Page Keeley Science Probe later in the lesson.

LESSON 2

Role of Animals' Eyes

ENCOUNTER
THE PHENOMENON

How do eyes help animals see?

GO ONLINE

Check out *The Eyes* to see the phenomenon in action.

Talk About It

Look at the photo showing different shapes and sizes of animal eyes. What do you observe? What questions do you have about animal eyes? Discuss your observations with a partner. Record your thoughts below.

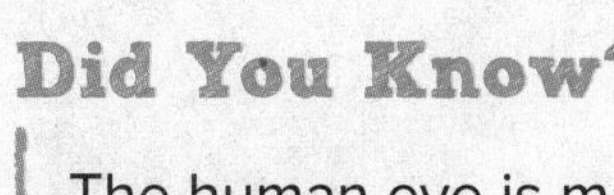
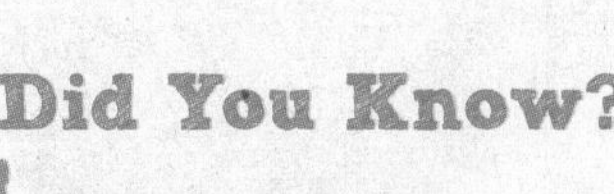

Did You Know?

The human eye is made up of more than two million working parts.

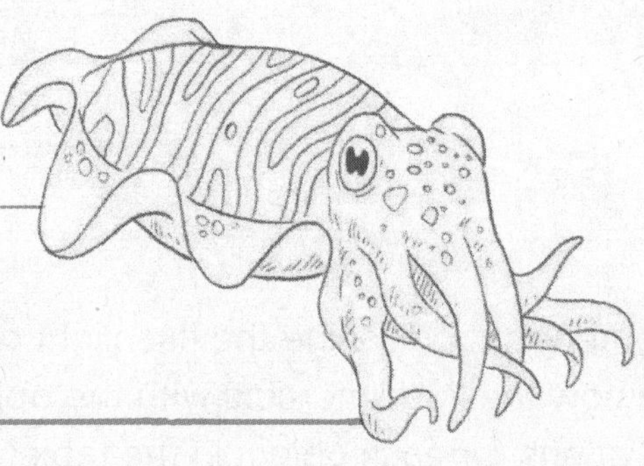

INQUIRY ACTIVITY

Hands On

How Light Travels

Animal eyes come in different shapes and sizes, but all eyes allow light to enter so objects can be seen. How does light from different objects get to your eyes?

Make a Prediction What happens when you shine a light on different kinds of objects in a darkened room?

____

____

____

____

____

Materials

mirror

white paper

flashlight

protractor

cup of sand

clear cup of water

index card

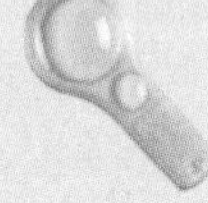
hand lens

Carry Out an Investigation

BE CAREFUL Do not shine the flashlight directly in someone's eyes.

1. Work with a partner. Use the mirror as a straightedge. Draw a line across the center of the white paper. Hold the mirror upright along this line.
2. Place the flashlight on the other side of the piece of paper so that it shines directly at the mirror.
3. Have your partner trace the ray of light. Place arrows to indicate direction.
4. Change the angle of the mirror. Have your partner draw new lines for the mirror and the ray of light. Use a protractor to measure the angle. What happened to the direction of the beam of light?

____

____

____

5. In a darkened room, shine the flashlight on each object, and record how the light interacts with the objects. Record your observations for each object in the table.

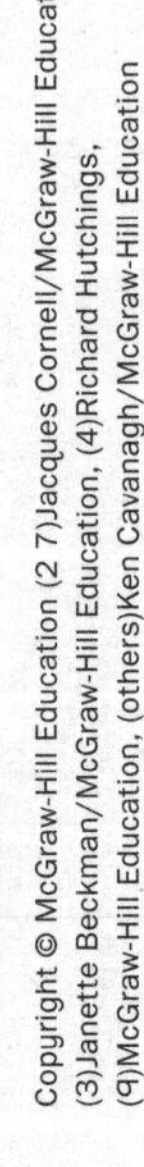

Communicate Information

Object	Observations
mirror	
sand	
water	
index card	
hand lens	

6. How does the surface of an object affect how much light bounces back?

7. How do you think the clear objects in this investigation affect the path of a beam of light?

How would you **use a model** to show how light travels? How would you **cause** the light to shift directions?

Talk About It

Do the results support your prediction? Discuss with a partner.

Reflection and Refraction

VOCABULARY

Look for these words as you read:

concave lens

convex lens

image

opaque

reflection

refraction

translucent

transparent

Recall how the beam of light traveled in the Inquiry Activity *How Light Travels*. Light has the properties of reflection and refraction.

Reflection is the bouncing of light waves off a surface. Most of the light that reaches your eye is reflected light. Look at your desk. If the desk did not reflect light, you could not see it. Most surfaces reflect at least some light. Smooth, shiny surfaces such as mirror reflect almost all of the light falling on them. Dull, rough surfaces reflect the least amount of light. The colors that you see are the colors that are reflected from objects.

When light reflects off a surface, it changes direction. Think about the mirror and flashlight in the Explore activity. The light rays moving toward a surface are the incoming rays. The reflected light rays are the outgoing rays. The angles of the incoming and outgoing rays are always equal. This is called the law of reflection. The **image** you see in the mirror is a "picture" of the light source that light rays make when bounced off a polished, shiny surface.

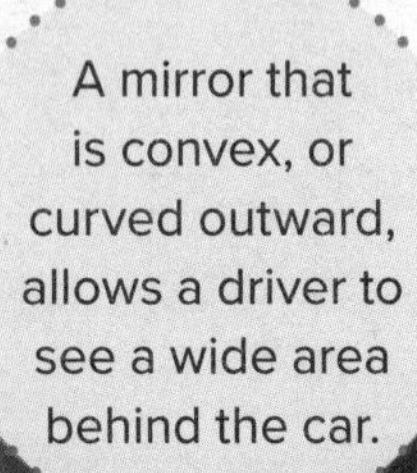

A mirror that is convex, or curved outward, allows a driver to see a wide area behind the car.

1. Draw a diagram to show how light allows objects to be seen. Include the light source, the eye, and label the direction of the light rays. Use the diagram to describe what happens if the eye is closed. What happens if the light is blocked or its path is changed? What if the light source is removed?

Refraction is the bending of a light wave as it changes angles passing from one substance into another. Light slows down when it moves from one material to a denser material. This decrease in speed causes the light's angle to change, or its direction to bend.

A clear piece of glass or plastic through which light travels is called a lens. If the lens is flat, the light's path shifts a little, but its final direction does not change. A lens that is thinner in the middle is a **concave lens**. Light that passes through a concave lens spreads outward. A lens that is thicker in the middle is a **convex lens**. Light that passes through a convex lens will come together at a focus point. The image you see when you look at an object through a convex lens depends on how far away the object is. Up close, the lens will cause the object to look bigger. Far away, the image will appear upside-down and smaller.

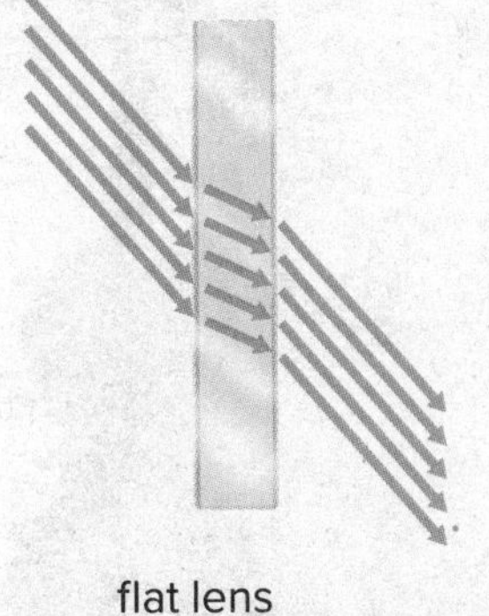

flat lens

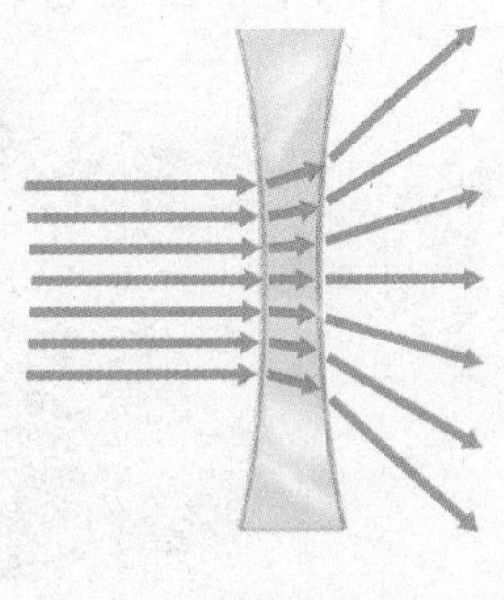

concave lens

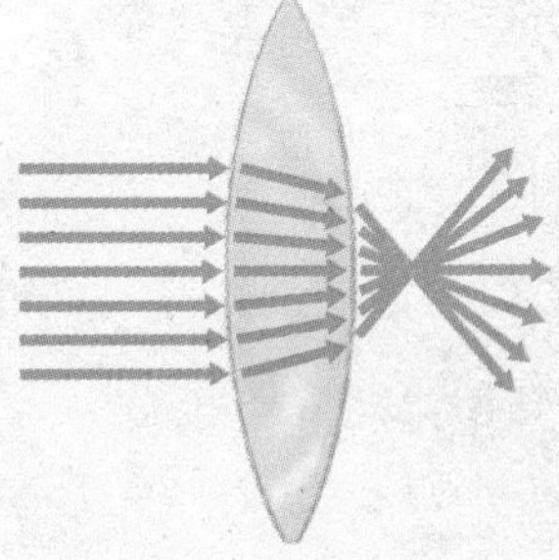

convex lens

2. What is the difference between refraction and reflection?

Concave lenses are used in eyeglasses for people who are nearsighted, or have trouble seeing objects that are far away. Convex lenses are needed for people who are farsighted, or have trouble seeing objects that are up close.

Talk About It

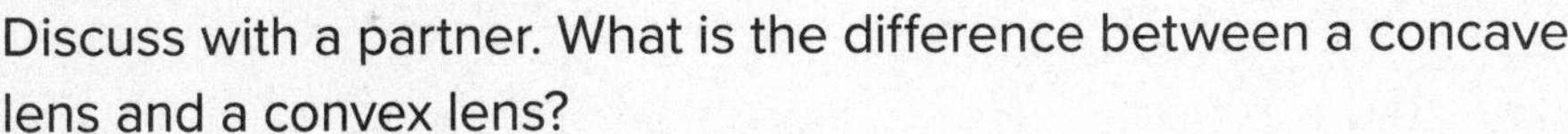

Discuss with a partner. What is the difference between a concave lens and a convex lens?

The Human Eye

You see an image when light reflects off an object and enters your eye. The diagram shows the different parts of the eye that light passes through. Light passes through the cornea and the pupil in your iris. The lens refracts the light so that it hits the retina on the back of the eye. The retina sends signals to the brain, and the brain interprets the signals as images. The eyes of other animals work in a similar way.

Read a Diagram

Trace the path of light as it enters the eye.

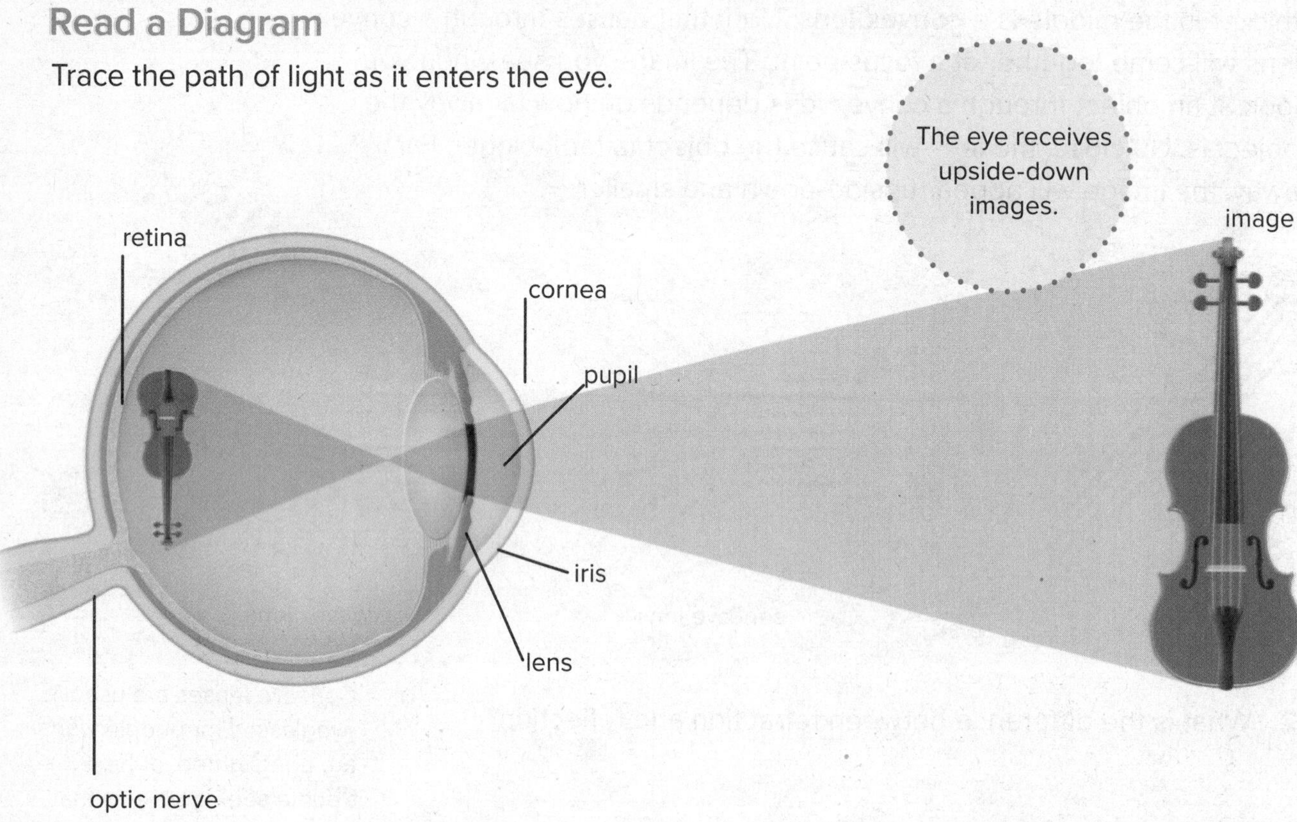

1. How do reflection and refraction allow animals to see?

____

____

____

____

____

REVISIT Revisit the Page Keeley Science Probe on page 73.

Transparent, Translucent, and Opaque

Clear glass lets light pass through it.

Frosted plastic scatters light in different directions.

Wood prevents light from getting through.

Light that strikes a material may pass through it. Materials that let light through, so objects on the other side can be seen clearly, are called **transparent**. Clear glass and clear plastic are transparent. You can see objects when you look at them through a transparent material.

Materials that let some light through, so objects on the other side appear blurry, are called **translucent**. Waxed paper and frosted glass are examples of translucent materials. If an object completely blocks light from passing through, it is called **opaque**. Whether an object is transparent, translucent, or opaque depends on the material, its thickness, and the color of light. Thicker objects tend to be more opaque.

2. **ENGINEERING Connection** When designing a new device, when would you want to use translucent materials? Give an example.

The Way Eyes See It

Eyes are like cameras. Both structures contain a lens that helps gather, focus, and transmit light. The brain uses information from the eyes to understand the world.

GO ONLINE Explore the *Animals' Eyes* simulation to change an animal's eyes and see how it affects its vision.

Animal eyes are different from human eyes. Different animals have different types of eyes, depending on where and how they live. All animal eyes use light in different ways to survive in the wild. Some animals have eyes on the front of their head, while some have eyes on the sides of their head. Still other animals, such as frogs, have eyes on the top of their head. Some animals can see many colors. Other animals can only see black, white, and grey. And some animals have better eyesight than others.

frog

fish

How are the eyes of animals similar to humans? How are they different?

chameleon

How Animals Use Their Eyes

The eyes of predators, animals that hunt other animals, are usually on the front of their head. This helps them to see how far away something is, especially when hunting for other animals. The eyes of prey, animals that predators eat, are usually on the sides of their heads. This helps them to see danger coming from the side and from behind them.

Lions have reflectors in their eyes that help them see at night. This makes them appear to glow.

Most fish have eyes on the sides of their heads. These eyes have special structures that allow fish to see movement underwater. This feature helps fish see creatures that may want to eat them. It also helps fish catch prey moving past them.

What can eyes located in front of the head do that eyes located on the sides of the head cannot do?

____

____

____

____

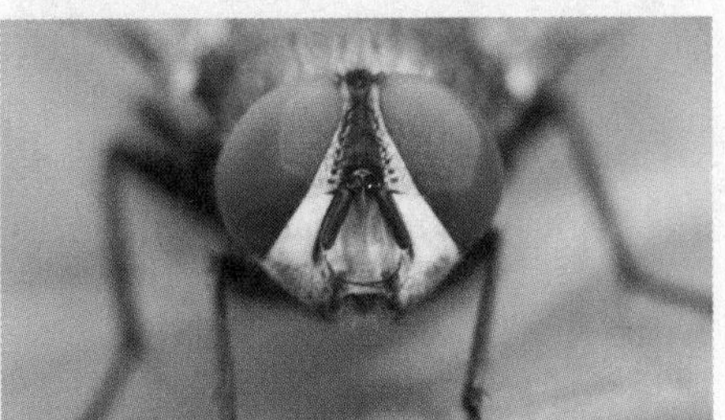

Flies, bees, and other insects have compound eyes that look like two big bubbles. Each eye can have thousands of small lenses. These lenses allow the insect to detect danger coming from all directions. This is why it is so hard to swat a fly!

Owls have eyes that are up to a hundred times more sensitive to low light than human eyes. This helps the owl to see prey at night. The eyes of an eagle are similar in size to human eyes. Objects appear much larger to eagles than they appear to humans. This helps eagles see prey from a distance. Eyes help animals in many different ways.

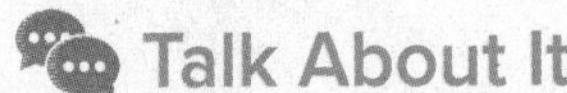

Talk About It

Compare the eyes of humans and birds. Discuss your ideas with a partner.

GO ONLINE Watch the video *How Do Animals See?* to learn more about how animals use their eyes.

INQUIRY ACTIVITY

Hands On

It's Time to Focus

Materials

- hand lens
- white piece of paper

Recall the characteristics of the front of your eyeballs. Humans and other mammals can focus by changing the shape of their lens. Fish and other aquatic animals live in a very different environment. Their eyes focus light in a different way. You will make a model to show how an aquatic animal's eye refracts light.

Make a Prediction What happens when you change the position of a hand lens between the white paper and the lamp?

Carry Out an Investigation

1. Hold the hand lens at arm's length between the lamp and the paper. What happens when you bring the hand lens closer to the paper?

2. Draw and label a diagram of your eye setup.

Communicate Information

3. Do the results of your investigation support your prediction?

How does your setup **model** the way a lens helps to focus an image? What **causes** the image to become blurry or crisp?

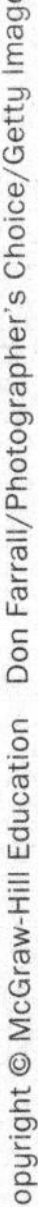

STEM Connection

What Does a Veterinarian Do?

Do you like to take care of animals? A **veterinarian** is a person who uses knowledge of animal structures and functions to help prevent, identify, and treat diseases in animals. These animal doctors help with many different types of medical issues. They also perform physical exams and surgeries. Treating eye problems is also an important part of a veterinarian's job. Animals, like humans, are prone to cataracts, which is an eye problem that can cause blindness if not treated. Veterinarians need to understand the structures and functions of animals' eyes to provide effective care.

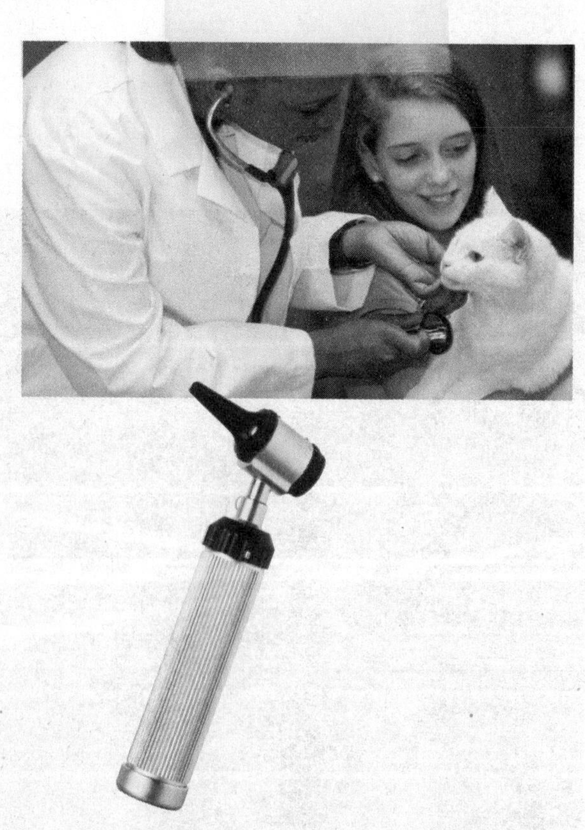

It's Your Turn

Think like a veterinarian. Complete the next activity to learn more about how eyes help animals survive.

Field of View

WRITING Connection The field of view is the area that you can see at any given moment. Animals and humans have a similar field of view. Research the advantages and disadvantages of forward-facing eyes and sideways-facing eyes. Construct an argument about the location of eyes that gives an animal a better chance of spotting predators. Write your explanatory paragraph on the lines below.

Inspect

Read the passage *Eye Technology.* Underline the text evidence that explains what it means to be nearsighted and farsighted.

Find Evidence

Reread the last paragraph. Highlight the text that tells about technology and cataracts.

Notes

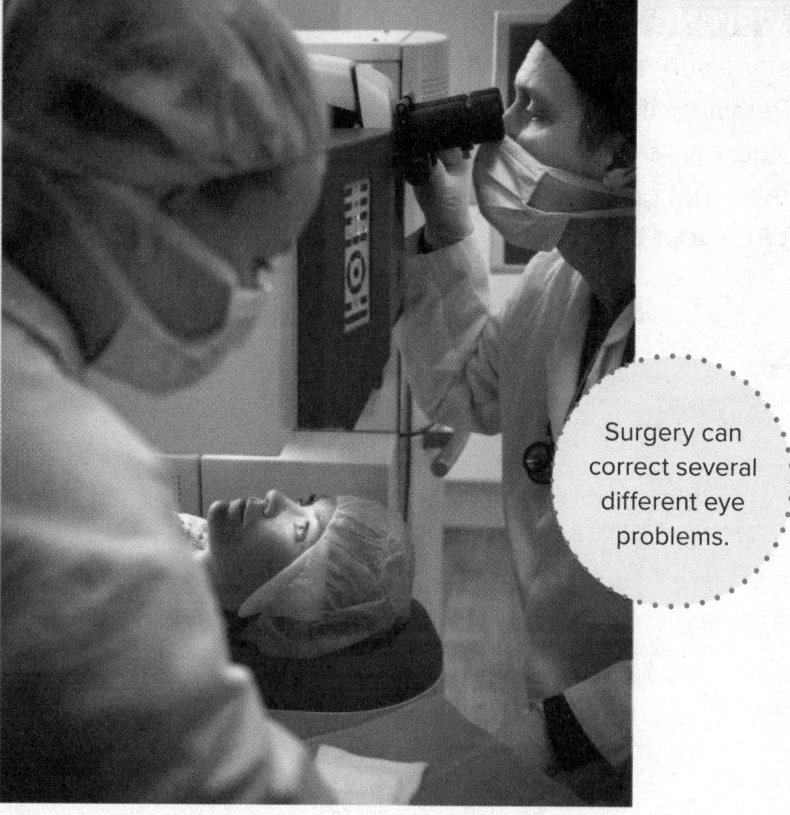

Surgery can correct several different eye problems.

Eye Technology

Our vision is very important to us. It tells us about our surroundings and warns us of danger. It also helps us gain knowledge through reading.

The eye responds to light. Light enters through the front of the eye. The cornea and lens focus the light onto the retina at the back of the eyeball. The lens is soft and flexible, so a ring of muscles around it can change its shape. When everything works right, vision is perfect. When everything does not work right, this can cause nearsightedness and farsightedness. Nearsighted people can see things up close, but distant objects look blurry. People who are farsighted can see distant objects but have difficulty reading and seeing up close.

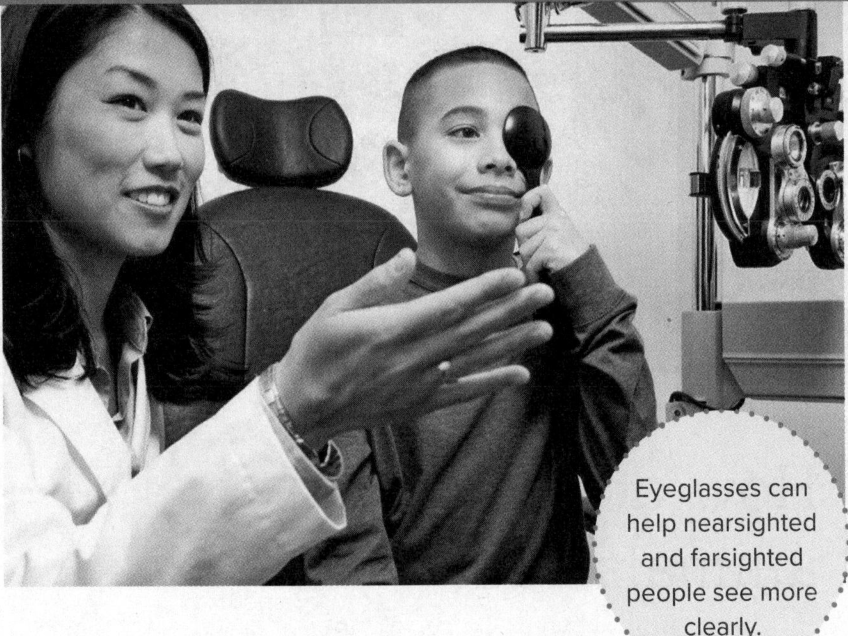

Eyeglasses can help nearsighted and farsighted people see more clearly.

Ophthalmologists, or eye doctors, can prescribe eyeglasses or contact lenses. The lenses used in eye glasses or contacts may be concave or convex. Concave lenses are for nearsighted eyes. They push the point of focus back towards the retina. Farsighted people need convex lenses to push the point of focus forward.

Cataracts cause the eye's lens to become cloudy. A person with cataracts has blurry vision. Colors are not bright, and lights may seem too bright or glaring. People with cataracts cannot correct their vision with eyeglasses or contacts. All is not lost, though. An ophthalmologist may be able to correct a person's vision with surgery. In the 1980s, Patricia Bath—a famous inventor—designed a tool called the Laserphaco Probe and a procedure to increase the precision of cataract removal. Her tool uses laser technology.

GO ONLINE Watch the video *The Eye* to learn more about how we see.

Make Connections

Talk About It

Look back at the STEM Connection of a veterinarian. Do you think animals have vision problems like those in this passage? Discuss with a partner how a veterinarian and ophthalmologist could work together to help animal's eyes.

Notes

LESSON 2

Review

EXPLAIN THE PHENOMENON | How do eyes help animals see?

Summarize It

Look at the photo again, and use what you have learned to explain the role of animals' eyes.

REVISIT PAGE KEELEY SCIENCE PROBES

Revisit the Page Keeley Science Probe on page 73. Has your thinking changed? If so, explain how it has changed.

Three-Dimensional Thinking

1. ______________ is the bending of light waves as they change angle passing from one substance into another.

   A. Reflection

   B. Refraction

   C. Translucent

   D. Retention

2. Most predators have eyes on the ______________ of their heads.

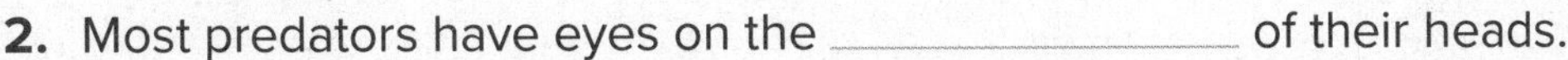

3. Explain how cataracts can have a negative effect on the role of animal eyes.

A beam of light inside a cave.

Extend It

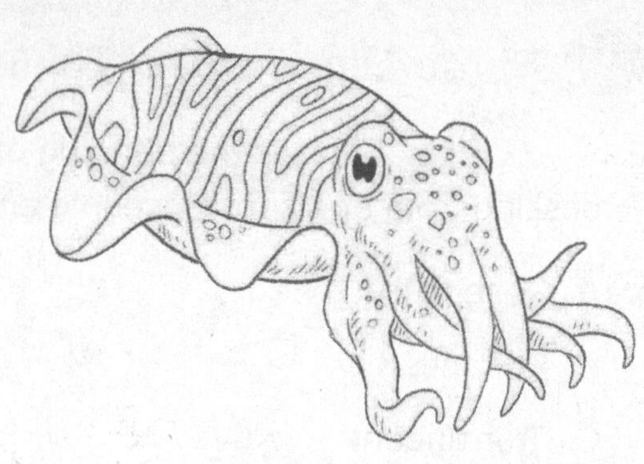

How are your eyes like a camera? Conduct research and develop a two-dimensional model comparing a human eye to a camera. To gather information, you can visit teacher-approved websites, read books at your local library, or interview an ophthalmologist. Include an explanatory paragraph with your model.

OPEN INQUIRY

What questions do you still have about how your eyes see?

Plan and carry out an investigation or research to find the answer to your question.

KEEP PLANNING

STEM Module Project

Engineering Challenge

Now that you have learned about animals' eyes, go to your Module Project to explain how this information will affect the design of your communication device.

LESSON 3 LAUNCH

Moving Information

Caleb and his friends wondered how information is moved from one computer to another. They each had different ideas. This is what they said:

Caleb: *I think the information travels as tiny bits of matter.*

Suki: *I think the information travels as waves.*

Rob: *I think the information travels as rays.*

Pita: *I think the information travels as electricity.*

Jean: *I think the information travels as sound.*

Who do you think has the best idea? ______________________________

Explain your thinking.

____

____

____

____

You will revisit the Page Keeley Science Probe later in the lesson.

LESSON 3

Information Transfer

ENCOUNTER
THE PHENOMENON

How do computers encode and transmit information?

GO ONLINE

Check out *Data* to see the phenomenon in action.

Look at the photo and watch the video of the computer code. What patterns can you observe? What do you wonder about computer code? Record or illustrate your thoughts below.

Did You Know?

Computers use patterns of zeros and ones to store data.

INQUIRY ACTIVITY

Hands On

Secret Message

Materials

Computers use patterns of zeros and ones, like those that you just saw, to communicate and store information. What other patterns or signals are used to help us communicate?

Make a Prediction How can you use light to send a secret message across a room?

Carry Out an Investigation

BE CAREFUL Do not shine the flashlight into anyone's eyes.

1. MATH Connection Work with a partner to develop a code of light signals for the alphabet. You will be sending a message to your group members on the other side of the classroom. Your message should not be able to be interpreted by your other classmates—only your group.
2. Go to the opposite side of the classroom from your group member.
3. Think of a one-word secret message that you will send to your group. Record your message below.

4. Send your message to your group using your code.

5. Switch roles within your group and repeat steps 2–4. Using your code, record the message your group member sends.

6. Use your code to translate the message.

Communicate Information

7. Did the results support your prediction? Explain your answer.

Compare your method of **transmitting the message** to the methods used by your classmates. Which worked best? What design criterion did you use to rate the codes?

Talk About It

How could you improve your methods or code?

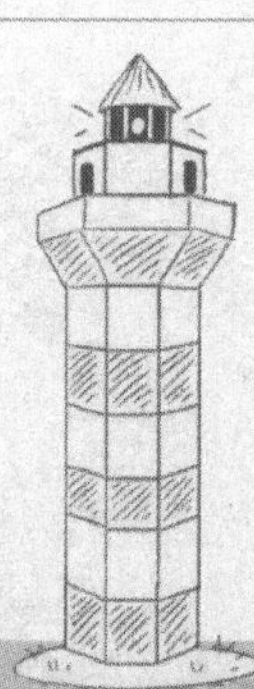

VOCABULARY

Look for these words as you read:

binary code

coding

Early Communication

Being able to sense and process information allows humans to communicate. We use our sight and hearing to sense and process messages. In the Inquiry Activity *Secret Message*, you used your sense of sight to communicate using patterns of light. Throughout history, engineers have found ways to use patterns to communicate. They use the design process to develop technology that makes the transfer of information faster and more efficient.

Before Electricity

Early communication methods included signal fires or smoke signals. Those methods required a clear line of sight so that the signals could be seen and passed on. Writing a letter and sending it with a messenger was another early form of communication. This method was very slow and sometimes unreliable.

Talk About It

What constraints would you face using smoke signals to communicate?

Smoke signals can be seen from miles away.

GO ONLINE Watch the video *Information Transfer* to learn about different ways information is transmitted from various sources.

Telegraph

The telegraph was invented in the 19th century. Solomon Brown, the first African-American to work for the Smithsonian Institution in Washington, D.C., worked together with Samuel Morse to revolutionize communication. The telegraph provided a faster and more reliable way to communicate over long distances. Information was sent through a wire using an electric signal. Telegraph operators would send electric signals in patterns of on-off tones, or clicks. This pattern system became known as Morse code. In the late 19th century, the telegraph was improved using wireless technology called radio.

Telegraph operators used patterns of clicks to send messages. They also interpreted these patterns and translated them back into messages.

Radio

Radio is the transmission of information using a certain wavelength of electromagnetic waves. Recall that waves are disturbances that transfer energy from one point to another. A radio wave is detected by a receiver and converted to a sound wave that humans can hear.

Sonar

Sonar is a system that uses sound waves to detect objects underwater. Some sonars send a sound wave, or pulse of sound. It then listens for the returning echo. The sound data is displayed on a monitor or heard on a loudspeaker.

How did people use patterns to send messages long ago?

INQUIRY ACTIVITY

Hands On

Morse Code Message

Materials

flashlight

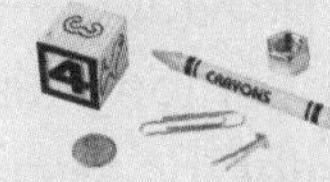
other classroom objects

In the past, using a telegraph to send a message by Morse code was a fast and efficient way to communicate over long distances. In this activity, you will investigate sending a message using Morse code.

Make a Prediction How can a device be used to send a message?

Carry Out an Investigation

1. Think of a short, simple message to send to your partner. Write your message below. Use the Morse Code, shown below, to code each letter of your message.

Letter	Code	Letter	Code	Letter	Code
A	.-	J	.---	S	...
B	-...	K	-.-	T	-
C	-.-.	L	.-..	U	..-
D	-..	M	--	V	...-
E	.	N	-.	W	.--
F	..-.	O	---	X	-..-
G	--.	P	.--.	Y	-.--
H	....	Q	--.-	Z	--..
I	..	R	.-.		

Your Message:

Your Coded Message:

2. Think of a way you can send your message using sound or light signals. Gather materials and construct any devices that you will need to send your message.

3. Send your message by making quick signals for a dot and long signals for a dash. Count 3 seconds between each letter and 5 seconds between each word.

4. As you flash the light, your partner will write the pattern in dots and dashes. Using the table, your partner will then decode the message.

5. Switch roles and repeat steps 1–4.

Partner's Coded Message:

____

Partner's Message:

____

Communicate Information

6. What challenges did you face in sending your message?

____

____

____

7. In what kinds of situations would Morse code be useful?

____

____

____

8. Did your results support your prediction? Explain your answer.

____

____

____

Advances in Communication Technology

Telephone

The telephone was an advancement of the radio system that allowed two or more users to conduct a conversation at the same time. The telephone converts sound into electronic signals that are transferred through cables. The signal is converted back into sound on the receiving end. In 1942, an Austrian-born movie actress, Hedy Lamarr, invented a device that would be the precursor to modern mobil communication technologies. Together with her business partner George Antheil, they were awarded the patent for a "secret communication system."

Cell Phones

Cell phones are a technology that allows people to communicate wirelessly. Cell phones are basically two-way radios. Each phone has a radio transmitter that sends a signal and a radio receiver that picks up signals. When a person talks on a cell phone, their phone converts their voice into an electric signal. This signal is then sent to the closest cell tower. The phone uses radio waves to send the signal. A network, or group, of cell towers then pass on the radio waves from one to another. Finally, the radio waves reach the other person's cell phone. The other phone changes the radio waves into an electric signal and then back into sound. This all happens instantaneously!

1. Draw arrows to show how signals are sent between the cell phone and cell tower.

Radio waves carry information from a cell phone to a cell tower and back.

Communications Satellites

Radio waves are also used by communications satellites. Communications satellites orbit Earth. These satellites can receive signals and then transmit them over long distances. Information is sent by radio waves to the satellite using a transmitting station. This is called an uplink. The uplink could be carrying phone calls, Internet information, or video information. Once the satellite receives the signal, it makes it stronger before transmitting it back to Earth. This is called a downlink. The re-transmitted signal is then picked up by a receiving station. Communications satellites allow information to be sent to places where cell phones or other methods of communication do not work.

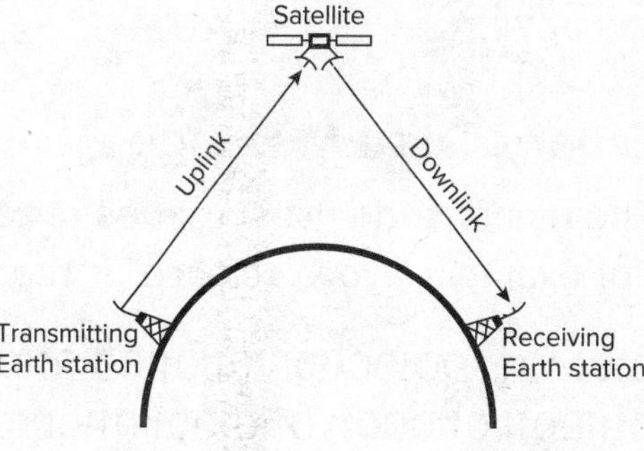

Over 1,000 working satellites currently orbit Earth.

2. What are the advantages of using a communications satellite?

Digital Communication

Technology has greatly improved our ability to communicate voice, text, images, and video over long distances. Computers and phones send this information using patterns. These devices can process binary code.

Binary code is a system that represents letters, digits, or other characters using zeros and ones. Using binary code, information can be sent quickly and accurately to another device. Photos, video, text, and voice information can be sent long distances. It can be coded and decoded without affecting the original information.

Computers use binary code to store and send information.

Binary Code Message

The binary code uses a series of zeros and ones to code a message. For example "I love science" is represented by the pattern below.

01001001 00100000 01101100 01101111 0110110 01100101 00100000 01110011 01100011 01101001 01100101 0110110 01100011 01100101 00100001

1. How is binary code similar to Morse code? How is it different?

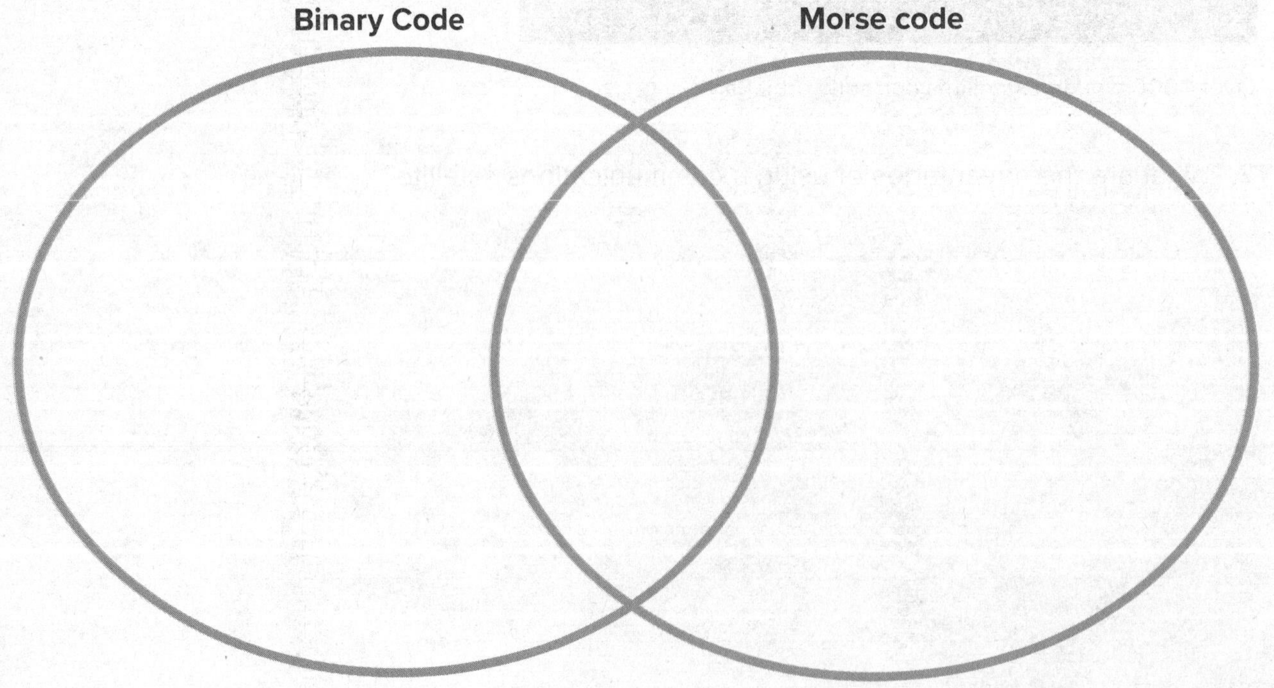

Sending Images

Technology can be used to communicate through images as well. Digital images are made up of many pixels. A pixel is the smallest piece of a digital image. Pixels are usually arranged in a grid. The pixels of different colors fill the rows and columns of the grid to form an image.

An image that is made up of few pixels will be grainy. An image that is made up of many pixels will be clear.

Coding

The process of writing a computer program in a language that can be used by a computer is called **coding**. Each line in a computer program gives the computer instructions for a different task. The languages are exact. They do not allow for errors or misinterpretation. The apps or programs on your cell phone or tablet were written using programming language.

Computer programmers use programming language to communicate with computers.

2. How have communication devices changed over time?

____

____

____

____

REVISIT Revisit the Page Keeley Science Probe on page 93.

PAGE KEELEY SCIENCE PROBES

STEM Connection

What Does a Computer Programmer Do?

Do you like to use computers? Are you good at writing instructions and solving problems? **Computer programmers** write the instructions that a computer needs in order to complete a task. Some tasks are pretty easy, like turning a light on or off. Other tasks are much more complex, such as tracking weather and making predictions.

Alan Turing (1912-1954), an English mathematician and logician, is known as the father of modern computing. He is credited with developing the ideas of artificial intelligence and of the modern computer.

It's Your Turn

Think like a computer programmer. Decode a binary code message in the next activity.

Talk About It

What would happen if there was an error in a computer code? Discuss with a partner.

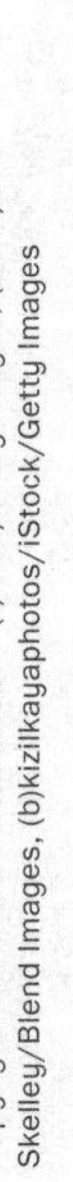

INQUIRY ACTIVITY

Research

What Does That Say?

You know that computers use binary code to communicate with other devices. You have seen messages represented in binary code and Morse code. How does binary code work?

Make a Prediction How can you code a message using only two numbers?

Take a look at the binary code below. This code represents a message that you will decode. Research the binary code for the alphabet in capital letters.

Binary Code Message:

01010011	01000101	01001110	01000100	01001001	01001110	01000111	00100000
01001001	01001110	01000110	01001111	01010010	01001101	01000001	01010100
01001001	01001111	01001110	00100000	01010111	01001001	01010100	01001000
00100000	01010111	01000001	01010110	01000101	01010011	00100000	01001001
01010011	00100000	01000011	01001111	01001111	01001100	00100000	01010011
01000011	01001001	01000101	01001110	01000011	01000101		

Carry Out an Investigation

1. Begin decoding the numbers, using what you have learned from your research.
2. Write the decoded message below:

Communicate Information

3. How is binary code used to send a message?

LESSON 3
Review

EXPLAIN THE PHENOMENON
How do computers encode and transmit information?

Summarize It

Explain how computers use patterns to send messages across distances.

REVISIT Revisit the Page Keeley Science Probe on page 93. Has your thinking changed? If so, explain how it has changed.

Three-Dimensional Thinking

1. A design criterion for an information transfer device is the accuracy of the message received.

Information Transfer Device	Accuracy of Received Messages (%)
Design 1	**89.8**
Design 2	**99.7**
Design 3	**96.4**
Design 4	**92.1**

The table above shows the results of a test on four different designs. Order the designs from best to worst.

2. Which of the following correctly represents how cell phones work?

   A. message → cell tower → phone 1 → phone 2 → cell tower → message

   B. cell tower → message → phone 1 → cell tower → phone 2 → message

   C. message → phone 1 → cell towers → phone 2 → message

   D. phone 1 → phone 2 → messages → cell towers

3. Which is not an example of using patterns to transfer information?

   A. Morse code

   B. binary code

   C. smoke signals

   D. thermal insulators

Extend It

Think about how communications technologies have changed over time. Research cell phones. Write a paragraph explaining how they have changed over the last 30 years. Predict how they will change over the next ten years. Use evidence from your research to support your prediction.

KEEP PLANNING

STEM Module Project

Engineering Challenge

Now that you have learned how information is transferred, go to your Module Project to explain how this information will affect your plan for your communication device.

Pixel Message

You have been hired as a telecommunications engineer to design a device that uses sound, light, or both to create two different binary codes. Once your device is ready, you will draw a pixel message on a 6x6 grid by shading some of the boxes to form a pattern. Then you will use the device to send the message across the classroom. Your group on the other side of the classroom will interpret the signals that you make with your device and replicate the pattern on their 6x6 grid. Be sure to repeat the procedure with the other binary code so that you can compare them and decide which way of communication was most efficient. You will use speed and accuracy as your criteria to rate your two binary codes.

Planning after Lesson 1

Apply what you have learned about information processing in animals to your project planning.

Draw a diagram that explains how internal and external structures are involved in sending and interpreting messages. Explain how this information will help you design a device to send an image.

Record information to help you plan your image-sending devices after each lesson.

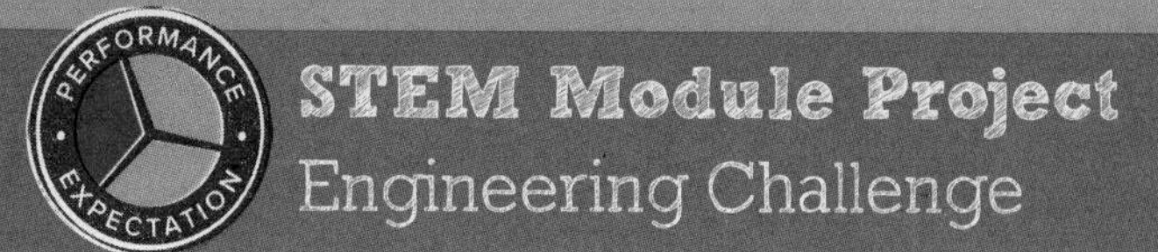

Planning after Lesson 2

Apply what you have learned about the role of animals' eyes to your project planning.

How do we see images? How will this information help you design a device to send an image?

Planning after Lesson 3

Apply what you have learned about information transfer to your project planning.

Use what you learned about using patterns to transfer information to develop the codes for your devices.

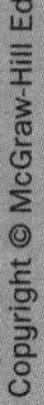

Define the Problem

Write the problem that you will solve in your own words. Identify the criteria and constraints.

Sketch Your Model

Draw your ideas below. Select the best two to build and test.

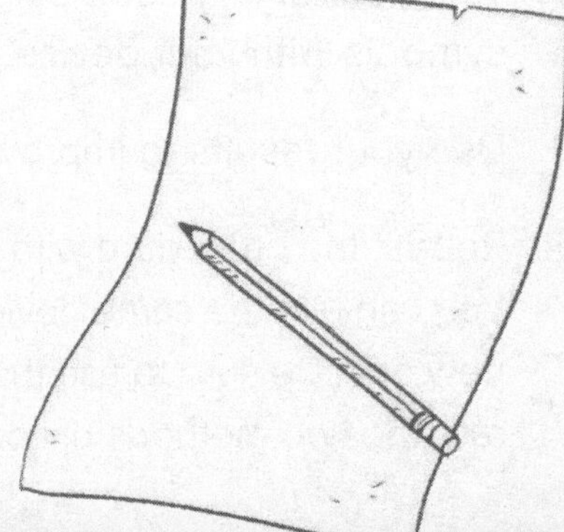

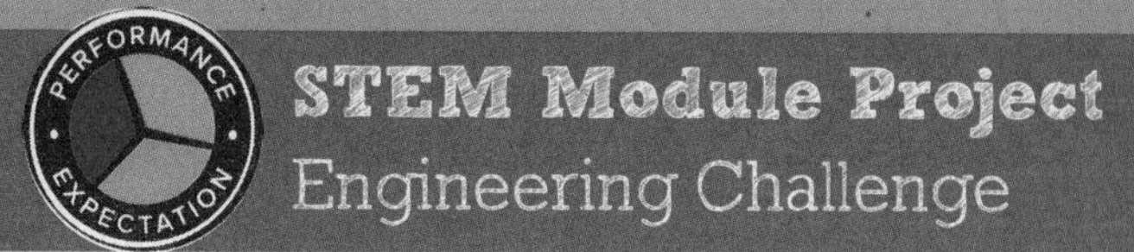

Pixel Message

Look back at the planning you did after each lesson.
Use that information to complete your final module project.

The Engineering Design Process

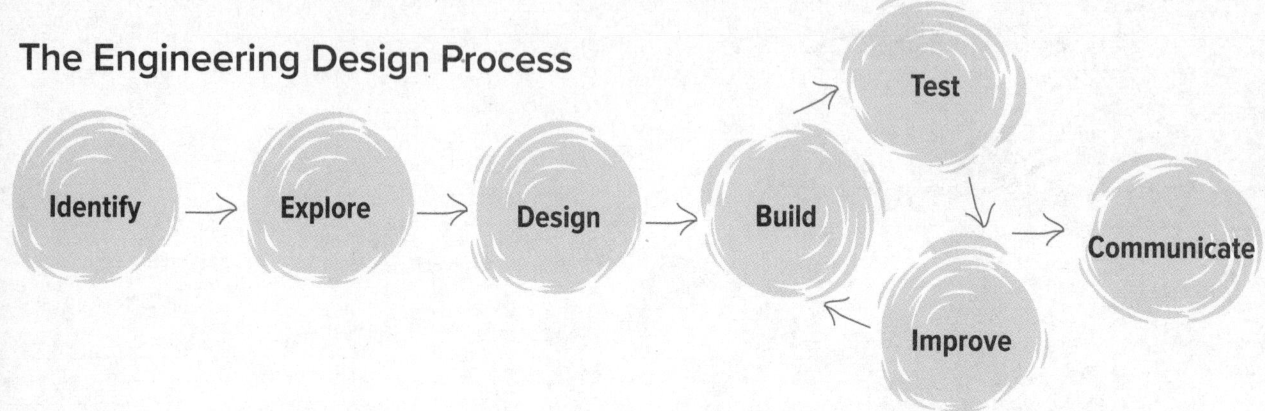

Build Your Model

1. Determine the materials that you will need to build your device. List the materials in the space provided.
2. Gather the materials needed. Remember that your device can use light, sound, or both to produce a two-symbol system of communication.
3. Use your project planning to create a binary code to communicate a message across the classroom.
4. Be sure that your group knows how to interpret the symbols or signals in your binary code.
5. Test your binary code by producing signals or symbols with your device.
6. Use your results to improve your model.
7. Repeat the procedure with a different binary code. You can use the same device, modify it, or build a new one. Be sure to use the project's criteria to rate the two methods of communication.

Materials

Procedure:

Test Your Model

Build your device(s) and test your two binary codes. Record the results of your tests below. Use a data table if you need to.

You are using the Engineering Design Process!

Communicate Your Results

Which of your two binary codes worked better? Share your codes, your devices, and the results of your tests with another group. Compare the ability of each of your devices and two-symbol codes to meet the criteria. Communicate your findings below.

MODULE WRAP-UP

REVISIT THE PHENOMENON

Using what you learned in this module, explain how the lighthouse transfers information.

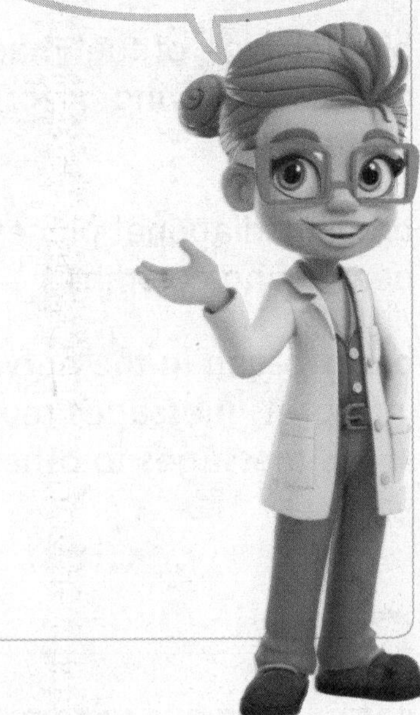

Have your ideas changed? Explain your answer.

Science Glossary

A

acceleration a change in velocity over time

adaptation a trait that helps a living thing survive in its environment

alternative energy source a source of energy other than the burning of a fossil fuel

amplitude a measure that relates to the amount of energy of a wave

B

binary code a system that represents letters, digits, or other characters using 0's and 1's

biofuel type of fuel made from biomass, or living or formerly living material

bracing diagonal pieces connecting beams and columns

brain organ in the nervous system that interprets messages received from and sends messages to other body organs

C

central nervous system part of the nervous system made up of the brain and spinal cord

chemical energy stored energy that is released when links between particles are broken or created

circuit a path through which electric current can flow

coding the process of writing a computer program in a language that can be used by a computer

collision when two or more objects crash into each other with a force

concave lens a lens that is thinner in the middle that always makes object look smaller

conduction the transfer of energy between two objects that are touching

conductor a material through which electricity flows easily

conservation the act of saving, protecting, or using resources wisely

conservation of energy a law that states that in a closed system, energy cannot be created or destroyed but can be transformed from one type into another

constraint something that limits or restricts someone or something

continent a large landmass

convection the transfer of energy in moving gases or liquid, such as warm air rising above a heater

convex lens a lens that is thicker in the middle that always makes object look larger

criteria standards on which a judgment or decision may be based

D

deposition the dropping off of eroded soil and bits of rock

design process a series of steps that engineers follow to come up with a solution to a problem

E

earthquake a sudden shaking of the rock that makes up Earth's crust

echolocation the process of finding an object by using reflected sound

electric current a flow of electricity through a conductor

energy the ability to do work

energy transfer the movement of energy from one object to another

erosion the movement of weathered material from one place to another.

external structures structures that are part of the outside of an organism's body

F

fault a break or crack in the rocks of Earth's crust where movement can take place

force a push or pull

fossil any remains or imprints of living things from the past

fossil fuel a source of energy made from the remains of ancient, once-living things

friction a force between surfaces that slows objects or stops them from moving

G

geothermal energy energy obtained from Earth's interior

H

heat the movement of energy from a warmer object to a cooler object

hydroelectricity electricity produced by waterpower

I

image a picture of the light source that light rays make in bouncing off a polished, shiny surface

inertia the tendency of an object in motion to stay in motion or of an object at rest to stay at rest

insulator a material that slows or stops the flow of energy, such as electricity or sound

internal structures structures that are found inside of an organism's body

L

landform a physical feature on Earth's surface

lateral force a force that comes from the sides

latitude the location north or south of the equator

longitude the location east or west from the Prime Meridian

longitudinal wave a wave vibrating in the same direction that the energy moves

M

magnitude the amount of energy released by an earthquake

medium a substance through which waves travel

motion a change in an object's position

N

natural resource something that is found in nature and is valuable to humans

nervous system the set of organs that use information from the senses to control all body systems

nonrenewable resource a natural material or source of energy that is useful to people and cannot be replaced easily

nuclear energy stored energy that is released when links between particles in the center of a particle of material are broken

O

opaque completely blocking light from passing through

P

peripheral nerve a nerve that is not part of the central nervous system and receives sensory information from cells in the body

plates large pieces that make up Earth's crust

pollution any harmful substance that affects Earth's land, air, or water

prototype an original or first model of something from which other forms are copied or developed

R

radiation energy that comes from a source in the form of waves or particles

reflection the bouncing of light waves off a surface

refraction the bending of light as it passes from one transparent material into another

renewable resource a useful material that is replaced quickly in nature

resistor an object that resists the flow of energy in an electrical circuit

response a reaction to a stimulus

S

sediment the particles of soil or rock that have been eroded and deposited

sedimentary rock a rock that forms when small bits of materials are pressed together in layers

seismic wave a vibration caused by an earthquake

seismograph an instrument used to detect and record earthquakes

sensory organ organs such as the skin, eyes, ears, nose, and tongue that gather information from outside the body

shear wall stiff wall made of braced panels

solar cell a device that uses light from the sun to produce electricity

solar power power obtained from solar energy to generate electricity using solar cells

sound wave a wave that transfers energy through material and spreads outward in all directions from a vibration

speed how fast an object's position changes over time at any given moment

spinal cord a thick bundle of nerves inside the spine

stimulus something in the environment that causes an action

structural adaptation an inherited change to physical features that helps an organism survive and reproduce

T

thermal energy the internal energy of an object due to the kinetic energy of its particles

topographic map a map that shows the elevation of an area of Earth's surface using contour lines

translucent letting only some light through, so objects on the other side appear blurry

transparent letting all the light through, so objects on the other side can be seen clearly

transpiration the release of water vapor, mainly through the small openings on the underside of leaves, that drives the movement of material throughout a plant

transverse wave a wave vibrating perpendicularly to the direction that the energy moves

tropism a plant's response to water, gravity, light, and touch

V

vegetation all the plants that cover a particular area

velocity the speed and direction of an object

vibration a back-and-forth motion

volcano an opening in Earth's surface where melted rock or gases are forced out

W

wavelength the distance from the top of one wave to the top of the next

weathering slow process that breaks materials into smaller pieces

Index

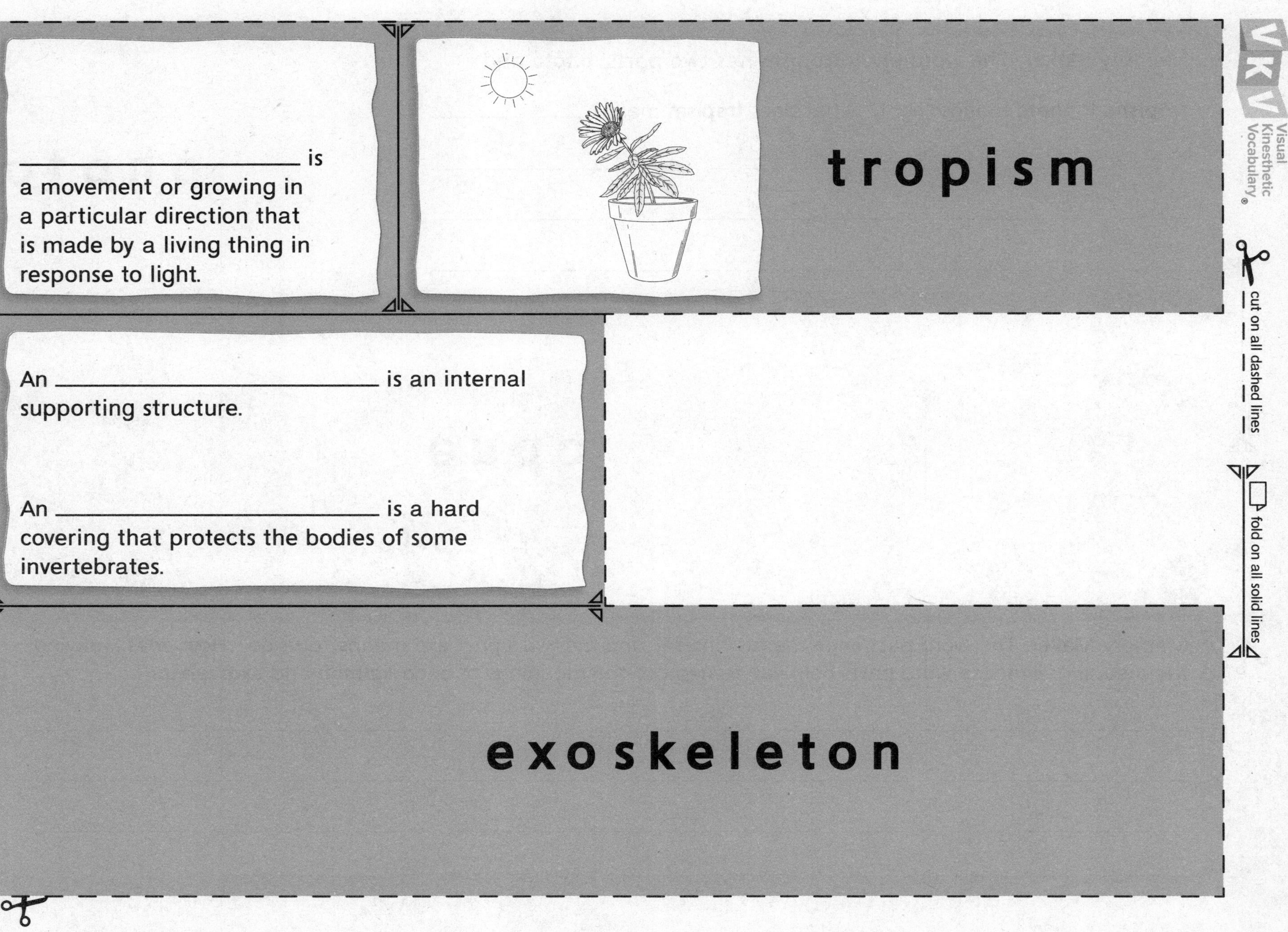

______ is a movement or growing in a particular direction that is made by a living thing in response to light.
tropism
An ______ is an internal supporting structure.
An ______ is a hard covering that protects the bodies of some invertebrates.
exoskeleton
Dinah Zike's
VKV
Visual Kinesthetic Vocabulary®
cut on all dashed lines
fold on all solid lines

Memory Maker: The word **phototropism** has two parts: **photo** and **tropism**. If **photo** means "light," what does **tropism** mean? ________

photo

endo

Memory Maker: The word part **endo** means "inside" and the word part **exo** means "outside." How does knowing the meanings of these word parts help you remember the meanings of **endoskeleton** and **exoskeleton**?

cut on all dashed lines

fold on all solid lines

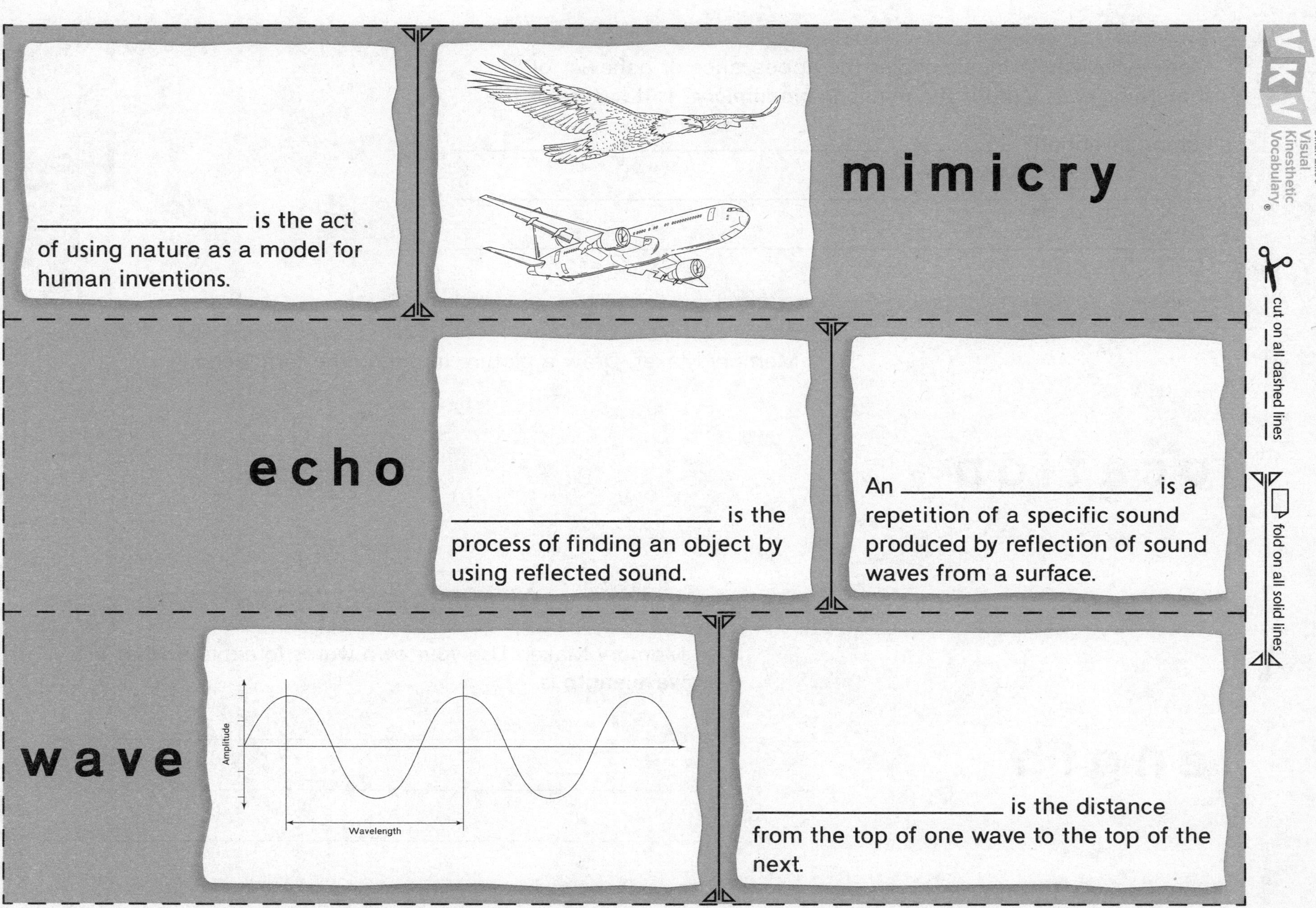

cut on all dashed lines

fold on all solid lines

Dinah Zike's Visual Kinesthetic Vocabulary®

cut on all dashed lines

fold on all solid lines

bio

Memory Maker: A mimic copies the appearance or behavior of something else. Who is the mimic in **biomimicry**? Is it nature or the human invention? ______________________________

____

____

location

Memory Maker: Draw a picture to define the term **echo**.

length

Memory Maker: Use your own words to explain what a **wavelength** is.

____

____

____